D0500378

Fly Me to the Moon

Belbruno, Edward, 1951-
Fly me to the moon : an
insider's guide to the n
c2007.
33305212222909
1a   05/17/07

Fly Me to the Moon

AN INSIDER'S GUIDE TO THE NEW
SCIENCE OF SPACE TRAVEL

Edward Belbruno

PRINCETON UNIVERSITY PRESS
PRINCETON AND OXFORD

Copyright © 2007 by Princeton University

Published by Princeton University Press, 41 William Street, Princeton
New Jersey

In the United Kingdom: Princeton University Press, 3 Market Place,
Woodstock, Oxfordshire OX20 1SY

All Rights Reserved

Library of Congress Cataloging-in-Publication Data

Belbruno, Edward, 1951–
    Fly me to the moon: an insider's guide to the new science of
    space travel / Edward Belbruno
        p. cm
    Includes bibliographical references and index.
    ISBN-13: 978-0-691-12822-1 (cl : alk. paper)
    ISBN-10: 0-691-12822-7 (cl : alk. paper)
    1. Gravity assist (Astrodynamics)—Popular works.   2. Celestial
mechanics—Popular works.   3. Chaotic behavior in systems—Popular
works.   4. Many-body problem—Popular works.   5. Outer space—
Exploration—Popular works.   I. Title.

TL1075.B45 2006
629.4'111—dc22   2006043749

British Library Cataloging-in-Publication Data is available

Printed on acid-free paper.∞

www.pupress.princeton.edu

Printed in the United States of America

10 9 8 7 6 5 4 3 2 1

This book is dedicated to the exploration of the Universe and to everyone who has helped me in my journey.

Contents . . . .

Foreword . . . .

Neil deGrasse Tyson

Sometimes, history comes fast.

A mere 65 years, 7 months, 3 days, 5 hours, and 43 minutes after Orville Wright left the ground on the first-ever powered flight, Neil Armstrong, the commander of *Apollo 11*, uttered his first comment from the Lunar surface, "Houston, Tranquility Base here. The Eagle has landed."

Of course the Wright Brothers' 1903 "aero" plane was heavier than air. Their new fangled machine would be quite useless across the quarter-million-mile airless gap between Earth and the Moon, as would every airplane invented and designed since. So the Apollo missions cannot be considered the natural extension of tinkering with winged craft, even though the lunar module of *Apollo 11* was named for a bird.

In the era before heavier-than-air flight, there was lighter-than-air flight—back when drifting slowly through Earth's atmosphere in the gondola of a hot air balloon was all the rage in the western world. To move effortlessly with the breeze was not enough for Jules Verne, the celebrated visionary and science fiction writer. Instead, he set the Moon in his sights, and knew that such an airless journey precluded the use of balloons and other imagined flying machines of the day. He proposed in his 1865 novel *From the Earth to the Moon* that this first trip, taken by three men, two dogs, and some chickens, would require a means of moving without the buoyancy of air. And so he loaded his spacefarers into a large bullet-shaped shell, and fired them out of an enormous gun; not unlike the circus performer who gets shot from a cannon to an awaiting net.

Verne badly underestimated the effect on the human body of explosively accelerating from zero miles per hour to Earth's escape velocity of seven miles per second. At the rate given, the three Moon voyagers and their animal companions would have become instantly pinned to the rear of the ship, and then crushed by the "g" forces into a pile of goo.

Carnage notwithstanding, the real message here is that, to Jules Verne and to the readers of his celebrated book, the Moon was not simply a cosmic object hovering at an unreachable distance. The Moon was as destination. The Moon was a world to be explored, with no less curiosity than one might carry to a distant mountain range, or to the far side of the ocean.

Curiosity. Humans have never lived without it. Who knows what thoughts course through the minds of other animals when they look up to the night sky. First, do they look up at all? And if they do, then do they wonder what's up there? Do any of them share our thoughts of adventure, as we stand on Earth's surface and feel the call of the cosmos? I doubt it. But even if they did, not only can we dream it, we can do something about it.

No, you can't (or rather, shouldn't) catapult complex living matter such as humans through space inside of oversized bullets fired through humongous cannons. You can, however, send people in rockets—an innovation that would await the creative efforts of American engineer Robert Goddard, who, in the 1920s pioneered liquid fuel propulsion. Upon the success of his rocket, especially the versions where you can throttle the thrust, thereby softening the g-forces on the occupants, Goddard immediately recognized the value of such a discovery for a Moon voyage, but was saddened by what would surely be the prohibitive cost of such a trip, lamenting that, "it might cost a million dollars."

Goddard was indeed a better engineer than economist: in modern dollars, that's about the cost of ten hour's time of the Space Shuttle in orbit.

@

Isaac Newton's law of gravity tells us that all cosmic objects pull on you at all times: the Sun, the Moon, Earth, other planets, the stars. And since everybody is in motion—in orbit around somebody else-your detailed trajectory through space can, and will be, quite complex. So, perhaps, we should not be surprised that people are still figuring out clever ways to maneuver around the solar system. One of my favorites is the gravitational slingshot. Today, with slim budgets, hardly any space probe has enough fuel to reach its destination without the help of another planet's gravity. The *Galileo* space probe to Jupiter, for example, required a three-planet gravity assist: first Venus, then Earth, and then Earth again, before heading for the outer solar system—a move that remains the envy of billiard enthusiasts.

Another of my favorites comes from a web of spooky, interconnected zones among the planets and their moons, where the sum of all forces is small, if not zero. Under these conditions, motion is practically effortless as you drift slowly through space. Barely understood until recently, and hardly explored until the work of Ed Belbruno, these newly discovered interplanetary highways offer a romantic reflection of the pre-rocket, pre-airplane era, where balloons would transport us, with hardly any energy of our own, from one unexplored vista to another.

Preface. . . . .

"Fly me to the Moon, let me play among the stars. Let me see what spring is like on Jupiter and Mars." Thus begins the song made famous by Frank Sinatra. Since the beginning of time, people have been gazing into space and wondering what might be out there. Each decade in recent history has witnessed a major event connected with space. Whether you were glued to your television set in 1969 to watch the first lunar landing, or recently, the Rover mission to Mars, there has long been a fascination with space and space travel. Many of us have wondered if we have neighbors in space or if we will be able to take a vacation to the Moon in our lifetime. Is it even possible? Not only is it possible, it is probable.

In 2004 Princeton University Press published my book *Capture Dynamics and Chaotic Motions in Celestial Mechanics: With Applications to the Construction of Low Energy Transfers*. After the publication of this technical monograph, I was pleasantly surprised to see the interest on the part of the media, which made a number of requests that I explain the content of the book to the more general reader. After encouragement from my editor, I wrote this little book on how the use of chaos is changing the way we maneuver in space. The 1987 publication of James Gleick's book *Chaos: Making a New Science* [30] introduced the general public to one of the most important

concepts involved in space flight. Chaos can be viewed as
a way to get a handle on the unpredictability in the sensitive motion of an object, such as a leaf falling through the
air, or a drunk trying to walk a straight line. In this book
the chaos results from the subtle intermingling of the gravitational pulls and tugs on an object moving in space. This
chaos can be utilized to obtain economical ways to open
up space flight to wider use and new applications. Many
interesting low energy and low cost paths in space can be
found. This approach is applicable to a whole host of interesting problems—not just controlling the motion of
spacecraft, but also understanding the motions of heavenly objects such as comets, asteroids, and even planets.

Using chaos to design new types of routes to the Moon
and other planets is a new and exciting development in
space travel. It was first applied in 1991 to rescue the lost
lunar mission of the Japanese spacecraft *Hiten*. The route
to the Moon used by *Hiten* was a substantial departure
from the classical route to the Moon—it used little or no
fuel.

The route to the Moon found for *Hiten* is just the tip
of the iceberg. The theory and method used to find it has
turned out to have many other applications for the motion of spacecraft. In 1998 I was lucky to be involved in
the salvage of another spacecraft. Like *Hiten*, this one,
called *HGS-1*, was never designed to go to the Moon, and
was actually a commercial Earth orbiting satellite that had
strayed into a nondesired orbit. The idea of using a lunar
transfer like *Hiten*'s, which led to a salvage of this satellite

by first bringing it to the Moon, was unique. In the early 1990s the European Space Agency was interested in using these methods for finding low energy lunar transfers, which led, in part, to its SMART 1 lunar mission. This spacecraft arrived at the Moon on its lengthy $1\frac{1}{2}$-year journey in November 2004.

These low energy routes can be used for purposes other than going to the Moon. In fact, they are being planned for future use by NASA for transfers between the various moons of Jupiter. Low energy routes also have interesting applications in the field of astronomy. The first concerns the motion of comets. The chaotic low energy route to the Moon is related to a special type of orbit for comets moving around the Sun. These elliptic-type orbits are synchronized with the motion of Jupiter, and the comets pass very close to Jupiter in their trek around the Sun. A comet in one of these orbits passing near Jupiter can be dramatically flung onto another totally different elliptic orbit around the Sun. However, the new elliptic orbit may be one that leads to the comet crossing the Earth's orbit and possibly colliding with it. One of these came very close to the Earth in 1778. These types of comet orbits have important implications for the future of the human race, since if a comet hit the Earth, most life would be destroyed. Kuiper belt objects, and other interesting topics, are all discussed in chapter 13.

Another application yields an intriguing theory on the origin of the Moon. The current accepted theory of the formation of the Moon is that approximately 4 billion years ago a Mars-sized planet slammed into the Earth, and the material that splattered out formed the Moon. But where did the mysterious Mars-sized world come from? This is the subject of chapter 14.

Chapter 15 includes applications of low energy chaotic transfers with missions to Pluto and beyond. But why stop at Pluto? The book ends with an interesting application of low energy transfers not just in our solar system, but much farther, with our closest stellar neighbor, the Alpha Centauri triple star system, which is four light years away. Although it is far away, and may seem to have little to do with our solar system, this need not be the case. This chapter reveals how comets can move from our solar system to the Alpha Centauri system.

You have read about solitary scientists working in their attics for decades to solve theoretical problems and the aha! moments that make all the pieces fall into place. There are also scientists who work on an idea for years, trying to convince colleagues of the practical applications. The implementation of a great idea must be done in collaboration with other scientists. This is a story of how that happens.

The goal of this book is to give a popular nontechnical treatment of how scientists determine space routes. The concluding chapter describes how the use of chaos to control the path of a spacecraft represents a paradigm shift in our way of thinking about space travel—about what is possible when we dare to think outside of the accepted norms.

Acknowledgments . . . .

I would first especially like to thank the National Aeronautics and Space Administration for their support, which has made this book possible. A special thanks to the AISR Program of NASA's Science Mission Directorate. NASA has always been a source of great inspiration in my life, both in my youth, remembering the *Apollo* landings on the Moon and the landings of the *Vikings* on Mars, and today with incredible visions of Saturn and our bold planned return to the Moon. I would also like to thank Princeton University, and, in particular, the Department of Astrophysical Sciences and the Program of Applied and Computational Mathematics. I am also very grateful for the support given to me by other institutions over the years, including the Geometry Center (formerly of the University of Minnesota), US Air Force Academy, McDonnell Douglas Corporation (now the Boeing Corporation), Alenia Spazio of Turino, University of Roma La Sapienza, the Planetary Society, Mitchell College, and New York Chapter of the National Space Society.

There are many people I have been fortunate to know and who I would like to thank: Joseph Bredekamp, Harley Thronson, David Folta, Robert Ceserone, Theodore Sweetser, James Burke, Daniel Goldin, Edgar T. Mitchell, S. Pete Worden, Edward T. Lu, Gordon Johnston, Elaine Walker, Candace Panakin, Samantha Surrey, Harold

Egeln, Eric Frydler, Roger Gilbertson, Taylor Dinerman, Michael Johnson, David Gump, Jane Barrash, Ester Gorsky, Robert Vanderbei, Richard Gott, Amaya Moro-Martin, Thomas McDonough, Louis Friedman, John Mather, Ingrid Daubechies, Donald Saari, Richard McGehee, Jaume Llibre, Ambassador Edward Finch, Claudio Maccone, Cesar Ocampo, Robert Bishop, Pini Gurfil, Giancarlo Genta, Senator Thomas Harkin, Michael Schulhof, Jon Schulhof, Irah Donner, Howard Marks, Rex Ridenoure, Herve Sainct, John Remo, Wendell Mendell, Brian Marsden, Jerrold Marsden, Michael D'Lorenzo, Douglas Kirkpatrick, Neil deGrasse Tyson, Victoria Sears, Richard and Janet Rose, Keith Gottschalk, Janet Dakey, Martin Hechler, Elbert Macau, Antonio Bertachini, Zhang Shiqing, Fillipo Graziani, Paolo Teofilatto, Gil Moore, Daniel Young, Robert Osserman, Barbara and Ellen Belbruno, and, in memory, Gerold Soffen, Carl Sagan, Juergen Moser, Herbert R. Shaw, E. Stamm, and S. Wolf.

The staff at Princeton University Press has been wonderful to work with throughout the entire process of writing and preparing this book for publication. First, I would like to thank Vickie Kearn, the editor for this book. Her great unwavering guidance and comments were invaluable in getting this book into final form. It was challenging for me to write a book of this type, and Vickie was always supportive as I fumbled through the early versions of the manuscript. I would also like to thank the production editor, Ellen Foos, who pulled everything together and made sure that I stayed on course. I am very grateful to

Dimitri Karetnikov for his great art direction and helping me get many of the figures in this book into final form. I would also like to thank the copy editor Marsha Kunin; Maia Reim for her wonderful art photography; and publicist Andrew DeSio. A special thanks to Lorraine Doneker for a great cover design. I would also like to thank all the other people at Princeton University Press who took part in putting this book together. Last, but not least, I would like to thank all the reviewers of the book manuscript, in its various stages, whose comments played a key role in getting the book into final form.

Fly Me to the Moon

Chapter One
A Moment of Discovery

. . . .

"Houston, we have a problem." That plea for help got Tom Hanks and his crew out of a jam on *Apollo 13*. But, who do you call when you don't work for NASA? . . . NASA!

At my door was a person I had never seen before. He introduced himself as James Miller. He had a problem.

The Japanese had launched a space probe to the Moon about three months earlier, in late January 1990. The main purpose of the mission was to demonstrate Japan's technical prowess in spaceflight. They had been gradually developing their technical abilities in space travel since the 1970s with less ambitious Earth orbiting missions. By 1990 they had built a considerable infrastructure to handle missions beyond Earth orbit including the Kagoshima Space Center. Now they wanted to become the first country to reach our neighbor after the Americans and Soviets. For Japan, this was an important mission, supported with national pride and a great deal of publicity.

But the mission had failed. Miller wanted to know: Could I save it? He had tried all the other obvious solutions and I was the last resort.

The Japanese had launched two robotic spacecraft MUSES-A & B into Earth orbit. These two spacecraft were attached to each other as they orbited the Earth. The smaller one, MUSES-B (renamed *Hagoromo*), the size of a grapefruit, detached on March 19 and went off to the Moon on a standard route, called a Hohmann transfer. But the Japanese lost contact with it, and it wasn't known if it ever made it to lunar orbit. It was last observed approaching the Moon, and preparing to go into orbit by firing its rocket engines, when communication was lost.

I was familiar with the mission, since it was widely broadcast in the press. One headline read, "Japan's Lunar Probe Lost." I didn't know much beyond what I heard through informal gossip from engineers in the hallways— that Japan was desperate to somehow get things back on track. The other spacecraft, MUSES-A, was renamed *Hiten*, meaning "A Buddhist angel that dances in heaven." Hoping to salvage the mission, Japan wanted to get *Hiten* to the Moon since *Hagoromo* appeared to be lost.

The Buddhist angel was the size of a desk, and was *never* designed to go to the Moon, but rather to remain in Earth orbit and be a communications relay for the now lost *Hagoromo*.

Miller was an aerospace engineer at NASA's Jet Propulsion Laboratory (JPL). He explained that he was trying to find ways for Japan to get *Hiten* to the Moon and into lunar orbit. But there were major problems—*Hiten* had very little fuel; it was not built to go to the Moon; and it would be impossible for it to reach the Moon by normal methods.

He asked if my theory of low fuel routes to the Moon could do it. He had heard that I had figured out a way to go to the Moon with much less fuel than conventional methods. He knew it was controversial, but was "willing to try anything."

I hadn't quite figured it out, but as soon as he asked me this question, it was like a light was turned on. As if the answer just jumped into my mind! I suggested that he do a computer simulation, and assume that *Hiten* was *already* at the Moon at the desired distance from it, and traveling with the right speed as specified from my theory. This was the first time I had ever applied my work to a real spacecraft, and there was no way to know if my suggested approach would be successful. The problem presented to me by Miller triggered the missing piece in my research that was needed to make my method work. It was one of those rare moments of scientific discovery that happen in the blink of an eye.

Miller was a bit skeptical that it would work. I gave him some initial critical parameters he would need to use in the computer simulation, and he left to try it out. I knew it was going to work.

He came by my office the next day, looking both excited and stunned—with computer output in hand, saying,

"It worked!" I was excited as well. Our results looked promising, but it would take some work to come up with a fully completed solution. So we started to determine a polished usable path to the Moon within the required margins. Not only would this path salvage the Japanese lunar mission, it would represent a new and revolutionary route to the Moon.

Chapter Two

An Uncertain Start

· ·

Big ideas start small. Some, as I've said, are aha! moments, others develop through years of research, and some, such as the *Hiten* mission, are a combination of the two. To see how we rescued the *Hiten* mission, we have to start at the beginning.

I arrived in Pasadena, California, in January 1985 to work at JPL. It was great to get out of the cold rainy climate of Boston, and the sunny, warm weather in Pasadena could not have been more welcoming. I was starting a new adventure in my life, but had no idea of the roller-coaster ride I was about to take.

I had just given up my position as an assistant professor of mathematics at Boston University. My research was not going as planned, and I was in need of some new ideas, often not easy in an academic setting in which one has to pay attention to what is acceptable or trendy at the time. So I felt JPL would be a perfect place to work. It sits in a small valley at the base of the imposing San Gabriel mountains, near Mt. Wilson.

JPL is responsible for exploring our solar system with robotic spacecraft. The list of their successes is many, notably in the mid-1960s the famous *Surveyor* landings on the Moon, and in the 1970s, the *Mariner* missions to Mercury and Venus, and the launch of the legendary *Voyager* spacecraft to Jupiter, Saturn, Uranus, Neptune, and beyond. Who could forget the historic landings of the two *Viking* spacecraft on Mars in the late 1970s. JPL's accomplishments are in the news today with the two Mars rovers, *Spirit* and *Opportunity*, rolling around on the red planet, and with the *Cassini* spacecraft arriving at Saturn with absolutely stunning views of Saturn's rings with its many moons—most notably the moon Titan. Titan's surface had been hidden from view by dense clouds. As this book is being written, Cassini, together with its *Huygens* probe, which landed on Titan, unveiled a bizarre world with possible oceans and rivers of liquid methane.

My new position was as trajectory mission analyst, and I was assigned to the *Galileo* mission to Jupiter. The *Galileo* spacecraft completed a successful mission on September 21, 2003, by plunging into Jupiter's atmosphere.

It was an eye-opening job. I had come from the very theoretical mathematical world of celestial mechanics where planets are regarded just as simple points. At JPL planets are naturally regarded as real objects. I had to get used to the fact that Jupiter was not a mere point anymore—that it has such things as a radius, a chemical composition, many satellites, magnetic fields, and radiation belts. Spacecraft were not just theoretical points either—they were real operational machines, with propulsion systems, guidance systems, communications antennae, computer operating systems, and so forth. My training was more concerned with the theoretical ways objects can move in space, and not focused on applications. The goal at JPL, on the other hand, was purely applied. They were interested in getting a spacecraft to a specific planet.

To make matters more challenging, I was used to the environment of academia where my colleagues were mainly mathematicians. Here at JPL, I was in a more company-structured environment, where most of my colleagues were aerospace engineers. They were less interested in the theory of how objects can move in space than in ensuring that a spacecraft successfully complete its mission.

This is a serious business—for several reasons. First, these missions are high profile and represent national aspirations and goals voted upon at the congressional level. They are also expensive—from hundreds of millions to billions of dollars apiece. For example, the cost of the *Cassini* mission was about $3 billion. So the bottom line here is that you don't want to make a mistake or take chances with these missions. Space travel and launching rockets into space is a risky business. Thus, the attitude at JPL is to be conservative in mission planning and to use tried and tested methods. This does not mean that engineers cannot be creative—but the creativity has to be carefully tested and monitored. My mathematical training at New York University's Courant Institute, under the direction of Juergen Moser, was more concerned with theoretical types of motion. When we talked, we used phrases like, "In general, it can be proven that," or "It is possible that the following motion will occur."

To a mathematician, the point is to show that a particular motion exists without paying too close attention to its precise explicit motion. The main issue is the *general type* of motion. Mission-design engineers are interested *only* in the explicit motion and not in hypothetical situations. They want to see it, measure it, and plot it. Thus, I encountered a sharp difference of cultures.

I felt like a fish out of water. As part of my job, I ran countless computer simulations of trajectories for the *Galileo* spacecraft from the Earth to Jupiter and compiled unending columns of numbers. It was tedious work. But it was bearable, since it supported the human race's quest to explore our solar system and get off of this planet. I was, however, beginning to question whether I had made the right decision in moving to California and giving up my academic career as a mathematician. The magnitude of the risk I had taken in switching careers was becoming more apparent.

Chapter Three
Conventional Way to the Moon

. . . .

I wanted to keep theoretical research active, and I had a gut feeling that I would find some promising new ideas at JPL from the way trajectories were designed for missions. One thing stood out and caught my interest. Spacecraft were launched to the Moon and other places of our solar system on Hohmann transfers. A transfer, in general, is a trajectory in space from one location to another.

The person who first formulated a precise way to determine special paths from the Earth to the Moon, or more generally from a given planet to another location in space, was the German Walter Hohmann in 1916. Hohmann transfers minimize the energy (fuel) required to go from one point to another in space under certain restrictive assumptions. Hohmann's theory for computing them is relatively simple, but effective in many situations.

Although it is applicable in a general setting, the best place to start is to describe a Hohmann transfer from the

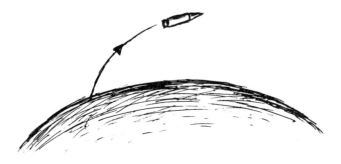

Figure 3.1 Rocket with payload going into Earth orbit

Earth to the Moon. Because there are many references on this, I'll just give a brief description of how it works. Let's start at the very beginning—on the ground.

A rocket lifts off from a launching pad on the Earth. It has several stages, each filled with the necessary fuel to reach orbit around the Earth. The uppermost stage contains a small payload—the cargo being carried into space. When this stage reaches Earth orbit, at a typical altitude of 124 miles (200 kilometers), the payload is released into circular orbit. It takes the rocket only a few minutes to reach this altitude, and all of the stages it used to reach this distance have fallen away as planned. Let's assume the payload is a small satellite planned to orbit the Moon, and it is now in a circular orbit 124 miles above the Earth. At the desired time, the satellite uses its own engines to give it the necessary kick to put it onto a path to the Moon (see figure 3.1). The moment these engines are started, the Hohmann transfer begins. The path to the Moon is fairly fast, taking

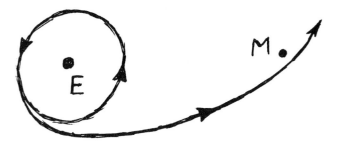

Figure 3.2 Hohmann transfer flying past the Moon

about three days, and when the satellite approaches the Moon at the desired altitude of, say 62 miles, (100 kilometers) from the lunar surface, it has to slow down by firing its rocket engines, otherwise it will just fly past the Moon (see figure 3.2). By the time the satellite, or spacecraft, reaches the Moon it is going about .62 miles per second (2,232 miles per hour) with respect to the Moon. If you want to slow down enough to place the satellite into a circular orbit around the Moon at 62 miles above the surface, you have to lose most of this high approach speed. You have to slow down by .56 miles per second (or 2,008 miles per hour)!

This circular lunar orbit is called a lunar capture orbit. If you look at figure 3.3, you will see that the engines are fired at two times. The first is when the spacecraft leaves the circular Earth orbit to go onto the Hohmann transfer to the Moon. The spacecraft needs to get a velocity boost of about 1.9 miles per second, which is quite a high velocity. (A bullet goes about .3 miles per second.)

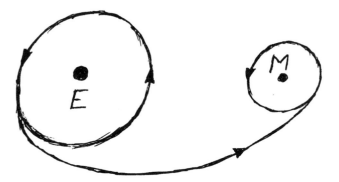

Figure 3.3 Hohmann transfer from circular Earth orbit to the Moon and into circular lunar orbit

This engine burn lasts only a few minutes and is called the Earth escape maneuver.[1] After this maneuver is done the engines are turned off, and the spacecraft then coasts to the Moon, which is about 250,000 miles from the Earth. When it reaches the desired distance to the lunar surface of 62 miles, the engines are fired again for a few minutes, this time to slow down and attain a circular lunar orbit. This is called a lunar capture maneuver. To slow down, the spacecraft has to turn around and fire the engines *opposite* the direction it is traveling. After this maneuver the spacecraft is orbiting the Moon. It may be necessary to do tiny engine firings on the way to the Moon to make minor course corrections. These are called orbit correction ma-

[1] In general, a maneuver is a change of velocity imparted to a spacecraft after firing its rocket engines.

neuvers. In this example, we are assuming, for simplicity, that there are no orbit correction maneuvers.

The Hohmann transfer is actually determined in a pretty straightforward manner. When the spacecraft leaves the Earth and goes to the Moon, the gravity of the Moon is ignored and the spacecraft is assumed to feel only the gravity of the Earth. This is because the Moon is relatively far away, and its gravity is not significantly felt by the spacecraft. This simplifies the problem to involve two bodies—the Earth and the spacecraft—and is called a "two-body problem." This is desirable since the two-body problem can be solved in a relatively easy way, and yield useful formulae.

Similarly when the spacecraft arrives at the Moon, the gravity of the Earth in relation to the spacecraft is ignored, and only the Moon's gravity is modeled since the Moon is now dominant. This results in another two-body problem, between the Moon and spacecraft. Ignoring the gravitational fields of bodies in this manner is not mathematically correct since the spacecraft feels the gravity of *both* the Earth and Moon at all times—but it turns out that these assumptions are sufficient for many applications. The simplified assumptions used by the Hohmann transfer are restrictive, and as we'll see a little later, eliminating them opens the door for new types of more efficient routes in space.

Hohmann was far ahead of his time, a true visionary. In 1925 he published a book on his work entitled *Die Erreichbarkeit der Himmelskorper (The Attainability of Celestial*

Bodies). This was based upon work he formulated about a decade earlier, shortly after the introduction into the marketplace of the commercial automobile, and the demonstration of airplanes, beginning with the Wright Brothers in 1903. Rocket flight beyond the atmosphere had not yet occured. The German V2 rocket would be the first to demonstrate this major milestone, but not until the 1930s. But Hohmann knew that this was coming. He knew that someday humans would venture out into space. He was already thinking about going to the Moon, Mars, Venus, and other planets in our solar system, and finding ways to do that. Starting in the 1960s his work was successfully implemented by the United States and the Soviet Union. It played a key role in the landing of the *Apollo* spacecraft on the Moon. Hohmann's work laid the foundations for the field of astrodynamics, which determines trajectories for spacecraft.

A Fuel Hog

The Hohmann transfer works well in practice, and most of NASA's deep space missions, including the *Surveyor, Mariner, Pioneer, Voyager, Galileo, Magellan,* and *Cassini,* have used them to explore the planets of our solar system. Why is that? First, these transfers are easy to understand and determine on the computer. This was especially true in the 1950s and 1960s when computers were not computationally too powerful. In fact, Hohmann transfers can even be determined with a pocket calculator. So, they are easily accessible. They formed the general basis for the trajectories and transfers of the NASA spacecraft and also those of other countries, notably the Soviet Union, from the 1950s onward. Second, they are reliable, and have worked successfully many times. Third, these transfers are relatively fast and get to their destinations in a sufficiently short amount of time. So, why change the methodology of orbit design?

The Hohmann transfer comes with a price and is expensive to use. When a spacecraft approaches the Moon, it is traveling at the high speed of 2,232 miles per hour. (Commercial jets go only about 500 miles per hour.) If you were on the Moon watching a spacecraft fly by on a Hohmann transfer, it would seem to race across the sky. It takes a lot of fuel, which is expensive, to slow down enough to go into lunar orbit.

Let's get a sense of just how expensive this can be. If, for example, we assume that the spacecraft has a mass of 2,200 pounds (1,000 kilograms), which is typical for a small robotic spacecraft, then about 528 pounds of fuel is required. Now, it is very expensive to bring one pound of anything to the Moon—about a quarter million dollars. So, this capture maneuver alone costs approximately \$132 million dollars! And you think filling up your car with gasoline is expensive? In this sense, Hohmann transfers are fuel hogs.

Not only is the Hohmann transfer expensive, it has other issues. A disadvantage of a Hohmann transfer is the risk inherent in its fast approach to the Moon, or other planetary destination. Upon arrival at the Moon the engines have to be turned on to slow down. However, the time to do this is very brief—only a matter of a few minutes. If, for any reason, the engines don't fire correctly during this time, the spacecraft will not be able to achieve lunar orbit and will fly past the Moon into space—and be lost. So, for a Hohmann transfer, the orbit capture maneuver is operationally critical and can lead to a total mission failure if not performed correctly. An example of this occurred on February 21, 1993, when the NASA mission to Mars, *Mars Observer*, unable to perform the capture maneuver at Mars due to a malfunction in the propulsion system, failed to achieve orbit, flew by the planet, and was lost.

If it were possible to transfer to the Moon and be captured in lunar orbit, where no fuel is required, this would dramatically reduce the cost of a mission to the Moon. But how could this be done?

such transfers existed, they had likely already been discovered by the scientists working in the United States and Soviet space programs. It was suggested I not waste my time on the problem. One person said that even if a ballistic capture transfer could be found, the transfer time to the Moon would probably be hundreds of years, and hence not of any practical value.

These quick negative responses indicated to me that the existence of ballistic capture transfers was not known, but I knew from a theoretical point of view, from the work of the Russian mathematicians V. Alekseev and K. Sitnikov in the 1960s, and of Juergen Moser in the 1970s, that these transfers might exist. And they demonstrated, in a theoretical setting, that it was possible for one body to be captured by two other bodies for all future time. This is called permanent capture and it occurs automatically if the conditions are right, so in this way it is analogous to ballistic capture. Even though their result was theoretical and didn't yield actual useful trajectories, it opened the door to the possible feasibility of ballistic capture transfers for spacecraft. In the 1960s Charles Conley, another mathematician, tried to find transfers from the Earth to the Moon that would arrive at the Moon with low velocity relative to the Moon. This would be like ballistic capture since the spacecraft would be going slow enough for the Moon to automatically capture it. Conley was able to show the existence of small segments of trajectories near the Moon that had near zero relative velocity, but he was not successful in finding trajectories from the Earth that connected up with them, though he continued to conjecture that it might be possible.

Aware of these results, I had a gut feeling that ballistic capture transfers existed, and that somehow the very subtle gravitational effects of the Sun and maybe other planets such as Venus, in addition to the Earth and Moon, could be used to find new types of transfers. This was all the more intriguing since the Hohmann transfer is modeled in a simplified way using the gravitational pulls between just the Earth and spacecraft or the Moon and spacecraft. It is known from the field of celestial mechanics, which studies the motions of objects under gravitational attraction, how complicated the dynamics of a moving spacecraft can get if the motion is modeled with the combined simultaneous gravitational pulls of several bodies on a spacecraft, say the Earth, Moon, Sun and other planets—especially if the trajectory has more time to reach the Moon than the very brief three days for a Hohmann transfer. Because of the speed of a Hohmann transfer, the more subtle gravitational pulls are insignificant. However, as it takes longer to reach the Moon, these subtle forces add up and can cause appreciable change. If more days of transfer time are allowed while all these gravitational tugs and pulls are subtly acting, the resulting motion could move in myriad ways as the spacecraft slowly meanders its way to the Moon—and certainly ballistic capture is possible. It seemed clear that the resulting motion would be very sensitive, and thus easily altered with small changes.

From this, it seemed that if ballistic capture transfers

existed, their motion would be sensitive, and thus chaotic in nature. This means that the motion could be significantly altered with the tiny changes discussed in more detail in the next chapter. The idea that ballistic capture transfers might be chaotic meant that their motion would be complicated, and difficult to study. The field of mathematics that studies sensitive chaotic-type motion is called dynamical systems, and it was new to the field of astrodynamics. So the use of dynamical systems represented a departure from the methodology of Hohmann. The work of Conley was a step in this direction.

But, although I felt ballistic capture transfers might exist, finding them was another matter. Even if they could be found, they would have to be applied to real spacecraft, which would introduce many constraints, one of them requiring a relatively short flight time. I set out to investigate this problem. It was slow going since I had to work on it outside of my regular and time-consuming duties designing trajectories for the *Galileo* mission to Jupiter.

Soon, though, NASA and the nation would be shaken up by the tragic loss of the space shuttle *Challenger* on January 28, 1986. In the days after this accident, we heard rumors that funding was to be cut at JPL. I knew I was vulnerable due to my theoretical mathematics background and would be one of the first to go. Soon thereafter I was transferred from *Galileo* to a new project that was a bit less mainstream, designing a lunar mission for a spacecraft that would be very small in mass, with unconventional rocket engines. And this turned out to be the serendipitous break

Figure 4.1 Get Away Special cannisters in the shuttle cargo bay

necessary for me to find a satisfactory solution to the problem of getting a spacecraft captured into orbit around the
Moon with no fuel. The spacecraft would fit into the payload cannisters in the shuttle cargo bay.

These cannisters, shown in figure 4.1, are relatively
inexpensive trash-can-sized containers designed to hold
various small payloads that are often used for university
experiments. They have lids that pop open, exposing the
payload to space. Designing a spacecraft of this type to go
to the Moon was challenging, as typically, a spacecraft
designed to orbit the Moon was the size of a small car.
Because the spacecraft was to fit into one of these Get
Away Special cannisters for low-cost payloads, the proposed
lunar mission was called *Lunar Get Away Special* (LGAS).

CHAPTER FOUR

Figure 4.2 Space shuttle lifting off

magazines—but never like this. There is a level of reality that can't be completely described and appreciated unless you are there. I was at the VIP viewing stands, with about five hundred other people. It was a carnival-like atmosphere. People talking, laughing. I saw little kids with signs saying, "Let's go, Dad!" I realized this was real, and looking at this huge machine off in the distance, I had the sense that it was also very dangerous. It was a balmy Florida morning, and you could hear crickets and frogs in the swamp in front of us. In front of the viewing stand was a huge American flag gently waving in the light breeze. Over the din of the crowd, via loud speakers, you could hear conversations between the shuttle crew and mission control. You had a sense of foreboding. This was not a TV program. At t minus 9 minutes there is a mandatory hold in the countdown, the opportunity to stop it in case there is a problem. Awaiting a verdict, the crowd grows suddenly very quiet. It is surreal. The announcer over the speakers instructs us that there is a go for launch. The crowd breaks into applause and cheers. You immediately feel the excitement that this rocket may actually take off very soon. Then, in the midst of this, the announcer instructs us to stand for the national anthem—most people are standing anyway. You feel tears welling up in your eyes as it is played. Some people are visibly crying. You could not help but feel proud. Then the searchlights go out in the distance. Our lights go off. It is pitch black. All you can hear are the frogs. It is silent, as if you were alone. Then, booming over the speakers: t-59, 58, . . . 10, 9, 8, 7,

6, 5, 4, 3, 2, 1. The entire horizon suddenly lights up. Night turns into day. It is blinding.

I have never seen such an enormous flash. The small lake in front of our viewing stand becomes visible, and the surface of the water is lit by dancing reflections of intense light. The light from the engines in the distance 3 miles away is so bright that you have to shade your eyes and turn your head away. It reminds me of the kind of intense momentary flash you see from a camera that makes you see spots for a few minutes after the picture is taken. However, this is not just from a tiny camera, but from the gargantuan engines, and it isn't momentary, but continuous. The sound is deafening and you almost have to cover your ears—some people do. There is a huge roar, and as the shuttle rises upward, you hear multiple rapid sonic booms—like the loudest explosions you would hear at the end of a fireworks show that shakes the ground. The air itself is shaking, and you feel vibrations in your chest. The air seems alive. I sensed an irrational fear that we were too close, but calmed myself realizing that many launches had been viewed from here. The air itself seems to heat up. It is simply awesome, and people all around you are crying out loud with tears streaming down their faces. The shuttle initially rises slowly like a giant technological masterpiece, a beacon of our desire to explore the universe; then it seems to very rapidly leap out of sight. It looks like a bright star moving up in the predawn sky, already hundreds of miles away after only several minutes.

The engines the shuttle uses are termed *high thrust*. They operate for short periods of time, usually a few minutes, causing a dramatic change in the velocity of the spacecraft. By comparison, the engines of *LGAS* are low-thrust ion engines. Ion engines work in the following way: A gas such as xenon is passed through high voltage to alter its atoms and produce a charge. These are called xenon ions and they are made so as to accelerate as they move out of an opening, creating a small thrust in the opposite direction.

Figure 4.3 shows that both the ion and chemical rocket engines work in a similar way. A chemical rocket explodes chemicals, creating a hot gas that exits the engine, creating a thrust. The ion engines use charged ions which exit the rocket engine. Each process moves the rocket in the opposite direction from the movement of the gases or ions. However, the thrust of a chemical rocket is stronger than the thrust of an ion rocket. This is because ions are very small, and as they exit the rocket, their small mass creates a tiny thrust. The thrust of the engines on *LGAS* is equivalent to the force felt by putting a dime in the palm of your hand. These engines have a thrust that is about *30,000 times* weaker than that of a chemical rocket! Although this would seem to imply that ion engines are far weaker than chemical rockets, they have one advantage in that they can be left on continuously. Their small thrust gradually adds more velocity to the spacecraft than chemical rockets can.

Chemical rockets can only be left on for relatively short bursts, and their fuel is used up pretty quickly. Ion

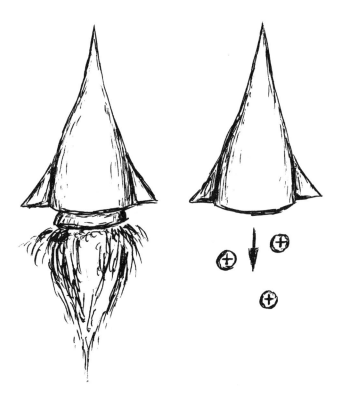

Figure 4.3 Thrusts of chemical and ion engines

engines use electricity to operate, which in the case of *LGAS* was generated by solar electric cells on long panels. Left on continuously, they can achieve a velocity change, called a ΔV, at any given moment, of only .001 feet per second. This is far less than the approximate 2,957 feet *per second* needed for lunar capture. A chemical rocket can achieve this in a *few minutes* of operation. It would take the engines of *LGAS* many days to achieve it.

My job on the *LGAS* project was to find a way to get *LGAS* captured into lunar orbit. Since it could only muster up a ΔV of .001 feet per second, it was clear that I had to find a way to get *LGAS* captured into lunar orbit where the velocity changes were almost zero (i.e., ballistic capture). This problem, which I was now working on serendipitously, turned out to be exactly the problem I was interested in and was researching. So now I had my chance to try and solve the ballistic capture problem for *LGAS*. But how? There was nothing about this in the literature.

Chapter Five
Chaos and Surfing the Gravitational Field

. . . .

Let's try to get an intuitive sense for how ballistic capture occurs. Imagine you are in a spacecraft approaching the Moon. If you approach too fast, the Moon's gravity won't be able to pull you into orbit. You'll just fly by it, like a Hohmann transfer. But if you approach too slowly and closely, the Moon's gravity will pull on you so strongly that you could crash onto the lunar surface (see figure 5.1).

So, the spacecraft has to slowly creep up on the Moon, like it is stalking it. It has to approximately match the Moon's circular orbit around the Earth. Since you would be approximately matching the Moon's velocity, then relative to an observer on the lunar surface, your spacecraft would appear to be almost motionless in the sky—as if it were hovering.

To accomplish this, the gravitational pulls of the Earth and the Moon on the moving spacecraft would have to

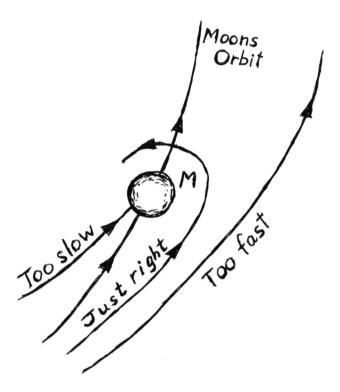

Figure 5.1 Sensitivity of ballistic capture

approximately balance. This type of balancing act is not too stable—sort of like someone on a tightrope. A slight increase or decrease in speed relative to the Moon is significant and would determine whether a spacecraft would fall away from the Moon or move sharply toward it.

This situation can be viewed as like a surfer riding the waves. If the surfer sees a good wave approaching, he must paddle to try and match the speed of the wave, stand up, and ride the crest. If he doesn't paddle fast enough at the right time, the wave passes him by. If he paddles too fast and too late, he falls off of the crest of the moving wave as he tries to stand up. Ballistic capture is similiarly sensitive to timing, and a spacecraft must thus arrive at the Moon with just the right speed. This type of sensitivity is called *chaos*.

What Is Chaos?

Let's define the term *chaos* as applied to the motion of an object. The motion of an object, which moves on a given trajectory, is called chaotic if a tiny change in the motion at some moment results in a large change of the motion of the object and a substantially different trajectory.

So, we can say that ballistic capture is a chaotic process and, as illustrated in figure 5.1, when it occurs, orbit around the Moon will be very unstable. It's like you're surfing the balanced gravitational fields between the Earth and Moon.

Chaos is an important concept, and it is used in many different contexts, so let's learn more about it. Consider a process of some sort, such as the motion of an object. It could be a falling leaf, or a spacecraft in motion, or a car. A particular motion of an object is chaotic if a tiny change in its motion causes infinitely possible changes in the direction of the motion. For example, have you seen a drunk man walking down the street? He doesn't really walk, but staggers along, weaving about. He has difficulty walking a straight line. If you give him a tiny push in a random direction, this change would likely cause a totally different path depending on the direction of the push. He could fall down, or move in any number of different ways. In this sense, a drunken walk is a chaotic motion, and in this example, one that is fairly robust and takes place rapidly.

Have you noticed a falling leaf on a windy day? Does it fall like a cannonball, straight down and hit the ground with a thud? Of course not. It floats in the air and moves in myriad possible directions depending on the wind. It flutters all over the place, and can go from one location to another in the blink of an eye. This too is a robust type of chaotic motion as shown in figure 5.2. Another example of chaotic motion is that of a glider plane as it gently moves on various air currents. A slight gust of wind can abruptly and unpredictably change the motion of the glider. These examples of chaos are pretty dramatic. Chaos can also be more subtle.

Consider the orbit of the Earth around the Sun. If we look at it in its entirety as in figure 5.3, it looks circular. There seems to be nothing chaotic about it. However, let's take a closer look and magnify a small part of this orbit.

If you look at figure 5.4, the circle is not smooth at all. It is irregular looking, with jagged curves. What the heck is going on here? Weren't we taught that the Earth goes around the Sun in an ellipse? Well, this is true only if we consider the gravitational fields of the Earth and the

Figure 5.2 Chaos of a falling leaf

Sun, where we model the Earth and Sun as points. We know from Newton's laws that the Earth does indeed move around the Sun in a perfect ellipse. On the other hand, if we include refinements to this model, then the Earth's orbit will be more complicated. For example, we could model the gravitational fields of all the planets, as well as the asteroids and the moons of the planets. The fact that the Earth, Sun, and planets are not points, but are spherical bodies with irregular shapes is a refinement. And refinements cause subtle changes in the orbit of the Earth and show up upon magnification.

These effects are observed in the modeling of the

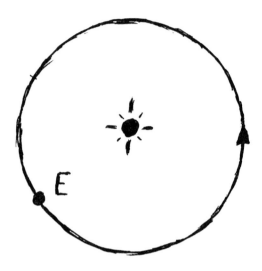

Figure 5.3 Earth's orbit around the Sun

Earth's motion using a planetary ephemeris, which is illustrated in figure 5.4. As mentioned above, it shows a piece of the Earth's orbit. (The ephemeris is a database and model that gives an accurate prediction of the motion of the Earth and other planets over long time spans.) Since the ephemeris takes into account the observed motions of the planets, it also reflects many effects other than gravity. These include, for example, the fact that the Sun is gaseous and not solid, and always changing its shape. It also reflects the fact that our solar system, consisting of the Sun and the known planets and other associated bodies, is itself moving around the center of mass of our

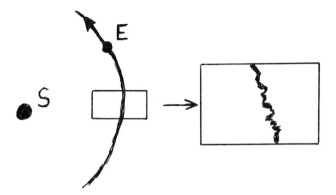

Figure 5.4 A tiny piece of the Earth's orbit

galaxy, and is gravitationally affected by it. In addition,
the gravitational effects of other stars, however small, are
also felt. There are also many forces of nature we either do
not understand (e.g., dark matter and energy) or of which
we are unaware. In sum, the motion of the Earth, when
magnified, is complicated. At each moment in time, it is
tugged and pulled, subtly, in infinitely many directions.
Unlike this subtle form of chaos, the process of ballistic
capture is more dramatic. We will read about it in the
next section. Later we will read about even more dramatic
types of chaos.

Chapter Six
Using Art to Find Chaotic Regions

· ·

When I agreed to work on the *LGAS* study in 1986, the study manager gave me only a few months to find a transfer to the Moon for the *LGAS* spacecraft using ballistic capture. Realizing that this was a chaotic type of process, and that it was not previously studied or in the literature, I knew that it was going to be challenging, with no guarantee of success. I needed any help I could get. I would also have to use the computer to illustrate this capture process. So where was I to start?

An Oil Painting Unveiling Dynamical Processes

Even before the computer was to be used, I needed to map out a region around the Moon where ballistic capture could occur. I knew this had to happen where the gravitational forces acting on the moving spacecraft would all tend to balance. But what did these regions look like? I found the solution by making a painting. I am an artist, and oil painting is my specialty. An artist I admire is van Gogh, and if you look at a van Gogh painting, you notice two things. The first is that you can see all the individual brush strokes moving all over the canvas. Second, what is striking about this, aside from the spectacular color, is the harmony. If you look very closely at a van Gogh painting, it looks chaotic. However, if you step back, all the brush-strokes settle into a scene with harmony, beauty, and splendor. An example most everyone knows is his painting *Starry Night*. Even though the night sky he painted appears to be explosive, there is a sense of harmony and balance to it. I sometimes use this type of effect in my work as well. The key is to paint very fast and spontaneously and unconsciously. Just let the painting emerge as the paint is applied intuitively to the canvas.

We see this, for example, in the painting displayed in figure 6.1 entitled *Dreams*. Now, how can this be used to find regions where ballistic capture can occur around the Moon? Art can be used to make insights into reality. So, if a painting is done of the Earth and Moon in space, and the brushstrokes are all displayed, they can be used to map out the approximate locations of the balancing of the intermingling gravitational fields of these two bodies.

This is illustrated in figure 6.2, which is the original picture I made to try and understand the balancing of the gravitational fields. Looking at this figure, you see that

Figure 6.1 *Dreams* (Edward Belbruno, oil on canvas, 36" × 48",
2003. Private collection)

around the Moon the brushstrokes, made with a pastel
crayon, are arranged in a circular pattern until they go a
sufficient distance, whereupon this pattern breaks down
and gradually transitions into the similar circular pattern
of the brushstrokes around the Earth. It is this transitional
region around the Moon that is of interest. It represents
the balancing of the two gravitational fields. If a spacecraft were moving in this transitional region around the
Moon, it would feel the gravitational tugs of the Earth and
Moon almost equally. It would be very weakly captured by

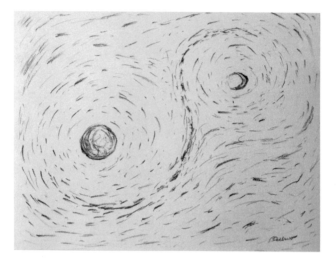

Figure 6.2 Picture of Earth-Moon system (Edward Belbruno, pastel on paper, 11" × 14", 1986)

the Moon, and would move about the Moon in a sensitive and chaotic manner. We call this transitional region around the Moon the weak stability boundary (WSB). It turns out to be a place where ballistic capture can occur, as we will see later.

Is there a way to visualize the WSB?

Chapter Seven
WSB—A Chaotic No-Man's-Land

. . . .

The weak stability boundary is a sort of "no-mans-land" around the Moon, because it is a place where the gravitational tugs from the Earth and the Moon are nearly balanced in relation to a moving spacecraft as it travels around the Moon. The spacecraft doesn't really know which body—the Earth or Moon—it is moving around; it is confused, so to speak. One moment the Earth pulls it away from the Moon, the next moment the Moon pulls it toward itself. This is a good situation for being captured around the Moon because while the spacecraft is in this region, the Moon is grabbing onto it very weakly, and so, it is *just barely* captured. This is what we sometimes refer to as *weak capture*, and it is analogous to a surfer riding the crest of a wave, as mentioned at the beginning of chapter 5. Again, the surfer is very delicately and chaotically balanced on a wave as he rides it. Likewise, a spacecraft is

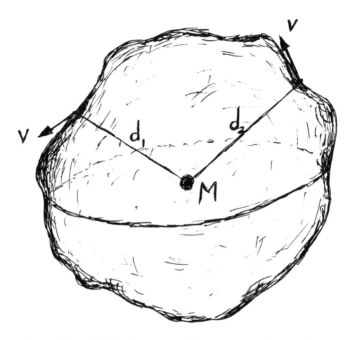

Figure 7.1 A WSB surface around the Moon specifying weak
capture for a given speed **v** at varying distances d_1, d_2

chaotically balanced as it moves in the weak stability
boundary.[1]

The WSB can be viewed in three dimensions as a set
of irregularly shaped surfaces surrounding the Moon. One
of these is seen in figure 7.1. This region can be close to the
surface of the Moon, in which case the spacecraft is moving

[1] Prior to 1990, the term *fuzzy boundary* was used in place of
weak stability boundary.

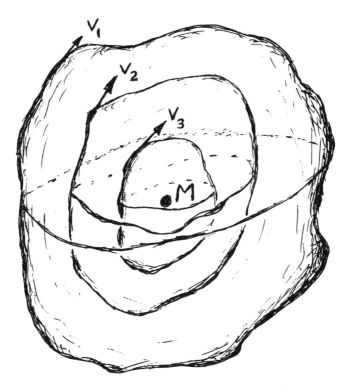

Figure 7.2 WSB surfaces surrounding the Moon for different speeds v_1, v_2, v_3

fairly fast, or it can be far from the Moon where the spacecraft is traveling much more slowly. You then get different WSB surfaces (see figure 7.2). The WSB is like an atlas.

Let's suppose you were approaching the Moon in your spacecraft and knew you had to arrive above the Moon at a certain fixed speed—over a particular location on the

Moon. If you wanted to be weakly captured with this speed, you would consult the WSB atlas and it would tell you how far above the Moon you would have to be for this to occur, depending on the latitude and longitude. This means that depending on your location over the Moon, you would have to be a particular distance above the surface from which your speed would just barely balance the Moon's gravity pulling you down, and the Earth's pulling you away. Only then could weak capture occur.

The chaotic nature of weak capture, is illustrated in figure 7.3. If you are going a little too fast you fly away from the Moon as seen in the trajectory marked (a). If your going a little too slowly, you get pulled toward the Moon shown in the trajectory marked (b). However, if you are traveling at just the right speed, you move weakly captured by the Moon shown in trajectory (c), where the motion is chaotic. An alternate and popular term for weak capture is *ballistic capture*.

It is possible to get another perspective of the WSB by using a coordinate system that rotates with the Moon around the Earth—a "merry-go-round" frame of reference where the Earth is at the center and the Moon is fixed on the moving platform. When you stand on the platform you will feel a third force—the pull away from the center, the centrifugal force. The WSB is the balancing of the gravitational forces of the Earth and Moon, together with the centrifugal force on a moving spacecraft.[2] An interesting

---

[2] It turns out that the WSB can be estimated by a computer algorithm, or with an explicit formula. See E. Belbruno, *Capture Dynamics and Chaotic Motions in Celestial Mechanics*.

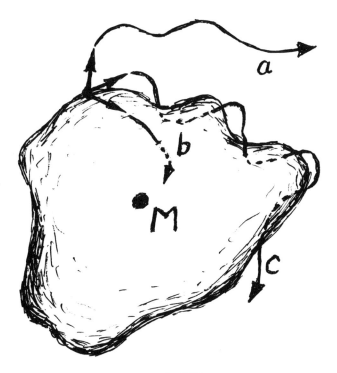

Figure 7.3 Chaotic motion in the WSB

case arises if the spacecraft is *not* moving in this system—
that is, if it is fixed relative to the Moon and Earth. Then
the balancing of these three forces occurs at just five
points. These are special points within the WSB when the
spacecraft doesn't move relative to the Moon. When it
does move, we don't get just five points, but rather the
WSB surfaces surrounding the Moon already discussed.
Now, when the spacecraft is fixed in this frame, these five
points are then five locations where all the forces are

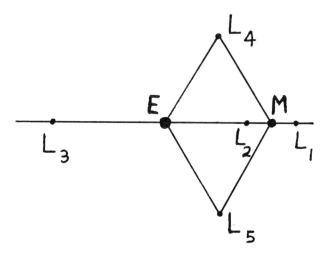

Figure 7.4 Lagrange points in a rotating frame of reference

exactly balanced. When a spacecraft is placed at one of these points, it will stay there forever. This is important in applications, since a spacecraft can be parked at one of these locations without any fuel for as along as desired.

These five locations have been known for a long time. They are called Euler-Lagrange points after the mathematicians Leonhard Euler and Joseph-Louis Lagrange who discovered them in the eighteenth and nineteenth centuries, respectively. They are labeled L_1, L_2, L_3, L_4, L_5, in figure 7.4. The points L_4, L_5 are the vertices of two equilateral triangles having the line between the Earth and Moon as the base. They are called equilateral points. The other three points lie on the line between the Earth and

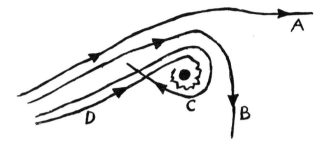

Figure 7.5 Evolution of gravity assist to weak capture by decreasing flyby speed

the Moon and are called collinear points. We'll come back to these points later.

The sensitivity of the motion of a spacecraft in the WSB can be viewed in another way. If a spacecraft approaches the Moon and flies by, its trajectory, labeled (A), is bent as shown in figure 7.5. In this case, the spacecraft will speed up as it flies by the Moon. It gets this increase of velocity because as it approaches the Moon, the Moon gravitationally grabs onto it. When it flies by, it gets a velocity increase due to the Moon's own motion around the Earth. This process is called gravity assist. If it flies by a little more slowly, the Moon has a longer time to grab onto the spacecraft, and the trajectory is bent more in the trajectory labeled (B). As the flyby speed is gradually reduced, the bending will gradually increase, until the spacecraft loops around the Moon as in trajectory (C), shown in figure 7.5. Eventually, as the flyby speed is further reduced, a special value of the velocity will occur where instead of looping around the Moon and leaving the Moon, the spacecraft will be moving slowly enough to be weakly captured, at the WSB, as in trajectory (D).

When a spacecraft is ballistically captured at the Moon, its motion is chaotic as we have seen. It might seem that it is not a great place to end up because of the sensitive motion. Fortunately this is not the case. For a tiny, though almost negligible maneuver, the weak capture can be stabilized into a lunar orbit that is not circular, but usually very eccentrically elliptic, and this is a remarkable property of ballistic capture that makes it ideal for applications.

Chapter Eight
Getting to the WSB—Low Energy Transfers

. . . .

You now know what a WSB is and some of its properties. Before we get back to the rescue of *Hiten*, let's explore the properties of the WSB to find a new type of transfer to the Moon.

We want to find a trajectory that goes to the Moon, and when it arrives at a desired distance from the Moon, that the spacecraft be ballistically captured. This means that the spacecraft arrives at the WSB (see figure 8.1). The term commonly used for this is ballistic capture transfer.

Returning to the *LGAS* study, I had computed the WSB of interest for the *LGAS* spacecraft so that we had a region from which ballistic capture could take place. This was the first part of the problem for the *LGAS* spacecraft.

Now that we had a region to go to, the next problem to solve was—how do we get there? This is not easy to do, because to start on a trajectory near the Earth, and have

Figure 8.1 Trajectory going to the WSB on a ballistic capture transfer to the location **a**

the spacecraft coast to the WSB with the desired velocity is like trying to thread a tiny needle. You have to give the spacecraft near the Earth a kick to send it on its way to the Moon, and after a number of days, or months, of travel it has to slowly creep up on the WSB with exactly the right speed. To make matters worse, the motion near the WSB is very sensitive and chaotic, so achieving the desired speed is a delicate problem. The way I tried to solve this problem was to very carefully adjust the size of the kick you give the spacecraft near the Earth. I found that this initial speed could not be adjusted to a fine enough degree to guarantee that by the time the spacecraft reached the Moon it would arrive at the speed required for the WSB. The spacecraft would either approach the Moon too quickly or too slowly. At the time I was working on this problem, tools had not yet been developed to tackle it. Was there another way to solve this problem?

The trick was to solve the problem in backward time. If you assume that the spacecraft is *already* near the Moon in the WSB with the desired velocity and distance, then

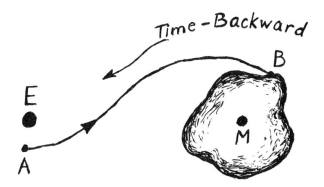

Figure 8.2 Finding a WSB transfer using the backward time method

you use this as the starting point. Next, you instruct the computer simulation to go backward in time and see where the spacecraft has to come from to get to your starting point. Hopefully, in backward time it will go near to the Earth. If it goes to a desired location, say A relative to the Earth, then in forward time we have a ballistic capture transfer from A to the WSB at the desired location, say B, with the correct velocity. This is called the backward time method. The properties of the WSB are ideal for this method since when in the WSB, motion is delicate and the spacecraft is weakly captured. It will stay in the WSB only for a very short time. So, in backward time the spacecraft will move away from the Moon, toward the Earth, and with a tiny adjustment of the velocity at B, you can make it go to A. This is illustrated in figure 8.2. The backward time method is similar to planning when you have to

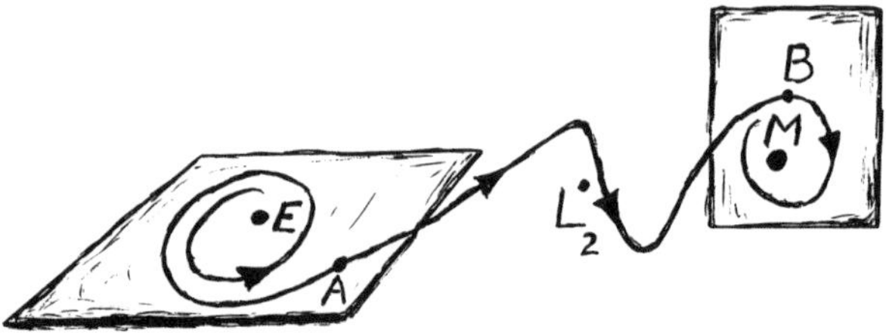

Figure 8.3 *LGAS* trajectory to the Moon

get up in the morning in order to get to work at a certain time and accomplish a number of things during the day.

When solving the *LGAS* transfer problem, it turns out that A is about 62,000 miles from the Earth, which is quite far. If it had more powerful chemical engines, it could quickly leave the Earth in a nearly straight-looking Hohmann transfer. However, since the ion engines are relatively weak, the spacecraft needs to slowly move away from the Earth, by gradually spiraling outward. The spacecraft took 3,000 spirals and 1.5 years to go from being released by the space shuttle to reaching A. At A the engines are shut off and the spacecraft coasts to B on a ballistic capture transfer, taking fourteen days. At B the engines are turned on and it spirals down to the desired operational lunar altitude of 62 miles in four months. So, the total time to the Moon after leaving the shuttle is about two years! This is a long time to reach the Moon, but the capture at B was ballistic—it required no fuel. You can see the total resulting trajectory in figure 8.3. As you will see, the European Space Agency *SMART 1* mission to the Moon, launched in 2003, was inspired by this *LGAS* design and confirmed the conjecture of Charles Conley we discussed in chapter 4.

A general ballistic capture transfer is called a low energy transfer since it uses no fuel to be captured into lunar orbit.[1] By comparison, as you will recall, a Hohmann transfer requires substantial fuel to slow down and be captured into lunar orbit and is called a high energy transfer since it gobbles up lots of fuel.

The existence of the ballistic capture transfer from A to B is artistically suggested in figure 6.2 where the transition between the circular brushstroke pattern around the Earth and the circular pattern around the Moon yields a path, shown by the darker brushstrokes, one could take to go from where the pattern around the Earth transitions into the pattern around the Moon.

[1] A ballistic capture transfer can alternatively be called a WSB transfer.

Chapter Nine

Rescue of a Lunar Mission

. . . .

OK, let's get back to the rescue of *Hiten*. After Miller appeared at my door in the spring of 1990, we found a way to get *Hiten* to the Moon following the same method used for *LGAS*. Remember, to determine the spacecraft's trajectory, we are assuming *Hiten* is already where you want to end up—near the Moon in the WSB. We are working backward in time with our model to see how it got there. The initial results looked very promising, but this type of rescue had never been done before.

Let's assume that B, in figure 9.1, is the starting point for *Hiten* in the WSB where it leaves the Moon in backward time using our simulation model. The results showed that *Hiten* had been pulled in the outward direction from the Moon, and from the Earth as well, as seen in figure 9.1. In fact, it went out to about four Earth-Moon distances from the Earth. It was far from both the Earth and the Moon, roughly 1 million miles from each. Then,

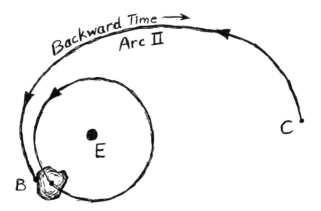

Figure 9.1 Trajectory arc, Arc II, from the lunar WSB to 1 million miles from the Earth using the backward method

continuing in backward time, it fell back toward Earth just as we had hoped. What we wanted was for the trajectory simulation to fall close to where the real spacecraft was located in its elliptic orbit around the Earth. In this way, the real spacecraft would be able to reach the Moon's weak stability boundary in forward time. Unfortunately, *Hiten* didn't fall as close to the spot near the Earth as we wanted, as is seen in figure 9.1. In this figure the motion of the actual spacecraft, not shown, is a little ellipse around the Earth, and our trajectory in backward time lies beyond this ellipse. What we had to do was find a way, using our model, to get the real spacecraft to somehow leave its elliptical orbit around the Earth in forward time, and join our trajectory in backward time at some location. It would then be able to find its way to the Moon's weak stability boundary.

The way we solved this was to first stop the backward trajectory at its farthest point C from the Earth, shown in Figure 9.1. Then, in forward time, this gave us a trajectory arc from C to the Moon, labeled Arc II.

That the trajectory moved so far out from the Earth turned out to be the key! The gravity of the Sun caused this by pulling the spacecraft away from the Earth-Moon system. (I had not modeled the gravitational effect of the Sun like this in previous work, so the trajectories didn't go too much beyond the Moon. I figured the Sun was too far away to have any appreciable effect.)

This discovery represented a real breakthrough! It was the key that provided the means to finding ballistic capture transfers to the Moon with the shorter flight times I had been looking for. The motion near C was very sensitive, and this provided the flexibility we needed. The reason the motion near C was sensitive is that it is near the balancing of the gravitational pulls of the Earth and Sun.

What we did next was to go back to where *Hiten* was actually in orbit around the Earth, as shown in figure 9.2. Our hope was that by applying a minimal maneuver to *Hiten*, it could escape the Earth and fly out to the point C in usual forward time. This maneuver had to be very small since *Hiten* had very little fuel available. After a little work, we found that a tiny maneuver could be found that would get us to C within our fuel budget. This was a relief!

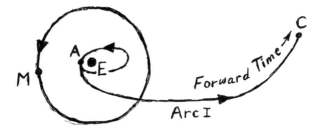

Figure 9.2 Trajectory arc, Arc I, from Earth orbit to 1 million miles from the Earth

We had another major issue to contend with. When this trajectory arc, Arc I, arrived at C, it had to match the velocity that the Arc II had when it left C to go to the lunar WSB in forward time. This required another maneuver to match the velocities. At first it seemed that the velocity mismatch was too large—needing substantially more fuel than was available. Initially I was skeptical we could accomplish this. Miller focused his time on this problem, and eventually, after many numerical simulations from the computer, we were able to lower the amount of fuel needed to pull this off. The flexibility of the motion of the spacecraft near C enabled this to happen.

This was exciting! We now had a trajectory, Arc I, that left from a point A on *Hiten*'s orbit around the Earth using a tiny maneuver, and then going to C, taking forty-five days. At C, *Hiten* had to do another tiny maneuver to go onto Arc II, then to lunar capture at B in the WSB one hundred days later. It turns out that the sum of the maneuvers to leave the Earth and the second maneuver at C was only 48 meters per second. This corresponds to using 6.6 pounds (3 kilograms) of fuel, which was hydrazine.

This was substantially less than the 100 meters per second, or equivalently 15.4 pounds (7 kilograms) of fuel, that *Hiten* had available! I previously would not have believed this was even possible. The complete ballistic capture transfer consisting of Arcs I, II from A to B is shown in figure 9.3.[1] It flew by the Moon after leaving the Earth to get a gravity assist.

We now had a viable route to the Moon for *Hiten*. The ability to achieve only 48 meters per second for *Hiten* to go from the Earth to lunar capture was substantially lower than the 250 meters per second that would have been required by a classical Hohmann transfer—way beyond the available fuel budget. Unlike the 3 days a Hohmann transfer needs to go to the Moon, however, this one took 150 days. Although this is a significant difference, it did not matter since it was a robotic spacecraft. The additional time also allowed us to monitor the health of the spacecraft.

We faxed this route for *Hiten* to Japan in June 1990. Frankly, I had little hope that the Japanese would pay too much attention to it. First, it was completely new and unlike anything that had been used previously for lunar transfers; and it used chaotic motion. There was nothing else in

---

[1] Unlike a Hohmann transfer, based upon modeling the motion of the spacecraft with either the Earth or the Moon in two-body problems, the ballistic capture transfer requires modeling simultaneously the Sun, Earth, Moon, and spacecraft—a four-body problem! Not too much is known about this problem.

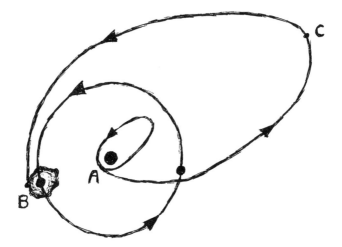

Figure 9.3 WSB transfer to the Moon found for *Hiten*

the literature about ballistic capture except that of *LGAS*.
Second, Miller and I were not working for the Institute for
Space and Astronautical Science (ISAS), the Japanese
space agency, and knew none of the people there.

To my surprise, they got back to us in only four days:
"We received your FAX dated on June 21. And thank you
for sending us an interesting paper. It is an amazing result
that a ballistic capture could be devised for *MUSES-A*.
We will reconfirm JPL results and send ours in a few days."
A few days later the Japanese confirmed that they repro-
duced the new transfer.

In the fall I left JPL and took a position at nearby
Pomona College. Being near JPL, I attended meetings with

the Japanese to discuss the new transfer to get *Hiten* to the Moon. It would be tracked by JPL's large Deep Space Network of radio telescopes situated around the world.

On April 24, 1991, *Hiten*'s engines ignited, and it was on its way to the Moon on the new transfer. The flight plan had it arriving in early October of that year!

When my position at Pomona College concluded in June 1991, I decided to resettle in St. Paul, Minnesota. At the end of September, I left Pasadena and drove to St. Paul. I would later hear, while driving cross country, that *Hiten* had successfully arrived at the Moon on October 2. *Hiten* is shown is figure 9.4.

There was more good news. The new transfer had saved so much fuel that it was decided to have *Hiten* slowly fly by the Moon upon arrival, and be only temporarily ballistically captured, and not be put into actual orbit as originally planned. It would then fly away from the Moon and travel within the Earth-Moon system for a few months performing more experiments in an extended mission.

The goal of the extended mission was to have *Hiten* fly near to the equilateral Lagrange points L_4, L_5 of the Earth-Moon system (see figure 7.4) and try to detect particles of matter. This is because L_4, L_5 are "stable" points. This means that any particles moving slowly near them would be trapped there and stay near them for long periods of time, maybe hundreds of millions of years. So, any matter that would be detected there might be very old and contain important information about the early solar system.

Figure 9.4 *Hiten* spacecraft, Institute of Space and Astronautical Science, Japan. Courtesy of the Planetary Society

Hiten successfully flew by these points and returned to the Moon in February 1992. No particles were detected at L_4, L_5, but this may be because *Hiten* was never designed to do this. It was placed into orbit with negligible fuel—it was approximately at the WSB. *Hiten* remained in lunar orbit for over a year. To end the mission, it was decided to crash *Hiten* onto the Moon as part of another experiment. The Apollo astronauts had placed seismic stations on the surface of the Moon. If these stations picked up the vibrations of the spacecraft upon impact, that would yield information about the structure of the Moon. *Hiten* crashed onto the surface of the Moon on April 10, 1993, ending an incredible space saga.[2]

Skepticism, Politics, and a Bittersweet Success

The road to the dramatic rescue of *Hiten* was not smooth, and many of my colleagues did not embrace the idea of using chaos theory, or methods of dynamical systems, to find new low energy routes in space.

This was first noticed in 1986 when the low energy transfer to the Moon was found for *LGAS*. Because ballistic capture was a new concept, and the *LGAS* mission itself was intriguing, it received some attention at JPL and in the press. A description of the ballistic capture transfer

[2] No vibrations were recorded at the seismic stations when *Hiten* crashed, likely due to the fact that the impact didn't have enough force.

for *LGAS* was first reported in the press in 1988 by the *Los Angeles Times* as part of a front page article, entitled "Cluster Probes Look for Lift on Space Guns". I was happy when it came out, but it quickly became clear that this *Times* story was not appreciated by some people at JPL. The approach my work took was different enough that its validity was questioned, which is natural. Even though the computer-generated output clearly demonstrated the validity of ballistic capture, it was not embraced by the engineering community. There are several reasons for this.

First, the idea of using substantially less fuel to achieve capture was not consistent with the way missions were planned, especially since the capture process was chaotic in nature. The term *chaos* implies a sense of unpredictability that is not consistent with the image of designing a set of predictable transfers, especially for expensive spacecraft. To make matters worse, the term *fuzzy boundary* had been used prior to *weak stability boundary* (see footnote 1, chapter 7). The earlier term had the sound of uncertainty, and was confused with fuzzy logic. Also, using less fuel could suggest a smaller spacecraft, which was contrary to the policy at that time of designing larger, more expensive spacecraft.

Second, my work used methods that were not familiar to the engineers. Because of the complexity of the capture dynamics, there was no way to easily give formulae to describe what I was doing. My methods were essentially numerical in nature, and the underlying theory at that time was not sufficiently well understood to explain it satisfactorily. So, it took on the air more of an art than of a science. Some engineers even referred to it as "black magic."

Third, the *LGAS* time of transfer to the Moon was about two years, and this sounded ridiculous to many people who were accustomed to the Hohmann transfer.

These issues got in the way of a proper appreciation of the potential applications of ballistic capture, and related interesting chaotic motions. Over the next several years I made attempts with little support at JPL to try and demonstrate that one could achieve much lower transit times for ballistic capture lunar transfer. With the assistance of a colleague from Barcelona, Spain, Jaume Llibre, I tried to obtain ballistic capture-type transfers to Jupiter that might have been applicable to the *Galileo* mission, but that too led to failure. My bosses lost confidence in my work.

Although I was convinced that I could reduce the transit time to the Moon, JPL was not. Because of these differences I decided in early 1990 to take a visiting teaching position at Pomona College in the fall, where I could also continue my research.

Serendipitously, very soon after I made my decision to leave JPL and go to Pomona College, I met Miller and was asked to find a way to get *Hiten* to the Moon in spring 1990.

This opportunity provided a way to actually demonstrate the validity of my theory. It was especially fortuitous given that only a few short months earlier, my work had been viewed as having little value, and I was let go from my job. Perhaps my work had a future after all.

Four years after the *Los Angeles Times* article appeared on *LGAS*, a second feature article was published, in June 1990 in the science section, announcing that there was a way to get *Hiten* to the Moon: "With a Boost from JPL, Japanese Lunar Mission May Get Back on Track." James Miller and I were thrilled to see this. In a few short months, work that my employer had viewed as having no useful applications produced a new type of transfer to the Moon that would actually be used. This naturally gave me a sense of vindication. However, this article was not well received by a number of people.

The day the article came out, I was driving around in my car listening to the radio, WABC, in Pasadena. It was just luck that I had it on, when the DJ was taking calls from people in reaction to the *Times* article. Their callers were not happy at all. I heard two of them, and they were critical of Miller, me, and JPL for "giving away American know-how and technology to Japan." They both cited what happened in the car industry as an example of how Japan used American technology to their advantage. It was amusing to hear this—but understandable.

Also, the Japanese at ISAS were not pleased with the article because they did not want to make it appear that their mission was being rescued by Americans. This is a sensitive issue in Japan, a country that sees itself developing its own independent space program. In fact, the Japanese wanted to make it appear as if *Hiten* had been *originally* scheduled to go to the Moon and as if *Hagoromo* had successfully orbited the Moon. The latter claim could not be made, of course, since communications were completely lost with *Hagoromo* prior to its possible arrival at the Moon and it was impossible at the time to view such a small object near the Moon. The fact that *Hiten* was only originally intended to be an Earth-orbiting communications-relay satellite for *Hagoromo* was well reported in the press.

Last, since I was being let go because my work was viewed as not having useful applications, the management of JPL would not welcome with open arms the successful use of my theory to rescue a lunar mission with a new type of transfer.

My feeling of success was somewhat dampened by these reactions, though these various political events were somewhat expected and illustrate what can occur in a well-established organizational framework when a radically new approach to doing things abruptly appears. However, as with all new discoveries, over time the system adapts especially when the usefulness of the discovery becomes apparent.

Chapter Ten

Significance of *Hiten*

. . . .

The route designed for *Hiten*'s new mission to the Moon represents a turning point in the design of lunar transfers, and transfers in general. A short history of the trajectory design of missions will put the significance of *Hiten* in perspective.

In the 1960s, missions to the Moon and other destinations were designed based upon Hohmann transfers, based on two-body modeling—between the spacecraft and the Earth, or the spacecraft and Moon, as we saw earlier. The resulting path to the Moon is just one half of a very thin ellipse that looks almost like a straight line (see figure 10.1). A departure from the Hohmann transfer was made in the late 1960s when *Apollo* went to the Moon. That was a free return trajectory. It looks similar to a Hohmann transfer on the way to the Moon, but when it arrives there it doesn't fly by as Hohmann's does if you don't slow down. It loops around the Moon as shown in

Figure 10.1 Hohmann transfer from the Earth to the Moon

figure 10.2, and returns to Earth. This is called a figure-eight trajectory. It was used by *Apollo* for safety reasons. If the astronauts have to abort the mission due to some catastrophic failure prior to reaching the Moon, as in *Apollo 13*, they would automatically return to the Earth. To find this route we don't do the modeling by breaking up the motion as two two-body problems, as with Hohmann, but model it as a *three-body problem* between the Earth, Moon, and spacecraft. So, the spacecraft feels the gravity of both the Earth and the Moon *together* at all times. This is more complicated than the Hohmann transfer and enables the free return to the Earth. The three-body problem cannot

Figure 10.2 Free return figure-eight lunar trajectory

be solved using explicit formulae as with the two-body problem.[1] We have to rely on the computer to model this problem and find the right trajectories.

The Hohmann transfer and the figure-eight trajectory have one thing in common—if you slightly alter your course on either one, they don't change too much, and their appearance looks the same. If a trajectory changes by a tiny amount when it is altered by a tiny amount, it is called stable. So, both the Hohmann transfer and the figure eight are stable, but they have one notable difference, as we just discussed—you can find explicit formulae to determine the Hohmann transfer, but you cannot do so for the figure-eight. In this case we say that the Hohmann transfer is solvable, and the figure-eight is nonsolvable.

Things get really interesting when you use trajectories that are unstable. A slight deviation to them causes the spacecraft to abruptly change its path to something that could look totally different, and to head in a totally new direction. Unstable trajectories will always be unsolvable and require at least three-body modeling.

The first mission to use unstable trajectories was the *ISEE-3 (International Sun Earth Explorer 3)*. It was a NASA mission launched in 1978, and the first spacecraft to fly to

[1] The three-body problem is a famous problem in mathematics, going back to the 1700s. It has challenged many mathematicians since then, including Euler, Lagrange, Poincaré, Kolmogorov, Moser. It has defied a complete understanding to this day.

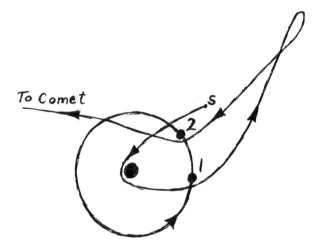

Figure 10.3 Part of the trajectory of *ISEE-3* starting at the location labeled **s**.

the L_2 point between the Earth and Sun, 930,000 miles from the Earth. It cycled around the L_2 point in unstable trajectories called halo orbits, based on the design by the NASA engineer Robert Farquhar. It was also the first spacecraft to be sent to a comet, Giacobini-Zinner, in 1985 from these halo orbits. It left the latter and traveled to the Moon, about a million miles away, and by using two lunar flybys (some very close to the lunar surface to within a few miles), it gained velocity via gravity assist to fly away from the Earth to reach the comet, shown in figure 10.3. The two lunar flybys are labeled 1, 2. It is not possible to show an entire trajectory plot for the *ISEE-3* mission due to its complexity.

The *ISEE-3* mission caught my interest in 1979 when I was in graduate school. I was amazed by the acrobatics of this spacecraft, as was my graduate adviser, Juergen Moser. This spacecraft could not have been flown using Hohmann-type trajectories due to the complexity of the motion. Since the flight of *ISEE-3*, there have been a number of other missions that have flown in halo orbits around the Earth-Sun Lagrange points L_1, L_2, such as WMAP (*Wilkinson Microwave Anisotropy Probe*), ACE (*Advanced Composition Explorer*), SOHO (*Solar and Heliospheric Observatory*), and *Genesis*.

The development of ballistic capture lunar transfers from 1986 to 1990, and, more importantly, their demonstration by *Hiten* in 1991, also represents the use of unstable trajectories. However, unlike the previous uses of unstable trajectories, for instance, by *ISEE-3*, the WSB transfers are designed by exploiting the unstable chaotic nature of the motion in a methodical manner, using principles of chaos. This is done for the purpose of dramatically reducing the fuel required—in this case, for the capture itself into lunar orbit. Thus, *Hiten* represents the first demonstration of a low energy transfer (defined in chapter 8). For the purpose of labeling, let's call *Hiten*'s transfer an exterior WSB transfer, since it moves way out beyond the orbit of the Moon. The one used by *LGAS* is an interior WSB transfer that stays inside the orbit of the Moon.

The WSB transfer used by *Hiten* also represents a fundamental new class of transfers that are substantially more fuel optimal than the Hohmann transfer. They also use a different type of capture process, which, unlike the Hohmann transfer, is gradual. This gradual capture process occurs for about two weeks prior to its arrival at its closest distance to the Moon. This means that the spacecraft can use smaller engines than with a Hohmann transfer to nudge it into lunar orbit. This would seem to imply that the spacecraft could be built much lighter than those using a Hohmann transfer. The Hohmann transfer needs large engines to quickly slow down the spacecraft within a few-minute time frame—or the spacecraft will be lost.

It turns out that the exterior WSB transfer can actually *double* the payload brought into circular lunar orbit as compared to a Hohmann transfer. This is because it needs about 25 percent less of a maneuver to place a spacecraft into low circular lunar orbit—at 100 kilometers altitude. This translates into using substantially less fuel than a Hohmann transfer, and this, together with using smaller engines, means that the mass of the spacecraft can be cut in half. Or equivalently, the saved mass can be used to double the payload. Since it costs about a quarter million dollars per pound to bring anything to lunar orbit, this is very promising for future lunar missions, and in particular, for constructing a lunar base.

Besides *Hiten* another spacecraft has used a WSB transfer. The European mission *SMART 1* arrived at the Moon in late 2004, on an interior WSB transfer inspired by the *LGAS* design. Japan's ISAS is planning a lunar mission called *Lunar-A*, which is also using the outer WSB transfer. This mission plans to send a robotic spacecraft to the Moon. It will fire penetrators into the lunar surface from orbit. The penetrators will analyze the lunar surface.

It is also possible to exploit the unstable chaotic motion near a halo orbit to dramatically reduce the fuel required to maintain the motion of a spacecraft in such an orbit. This was developed in 1985 by Spanish mathematicians Carles Simó and Jaume Llibre. Their methods were further developed by Caltech's Jerrold Marsden, together with Shane Ross and W. Koon; Martin Lo at JPL; and Kathleen Howell at Purdue University. Their trajectory design was demonstrated by NASA's *Genesis* mission in 2001.

J. Marsden et al. have studied the exterior WSB transfer and have gained some insights into its dynamics by exploring special pathways they travel on, which look like tubes. These pathways are part of what is sometimes referred to as the interplanetary superhighway. A number of other people have studied ballistic capture transfers from different perspectives, including M. Bello-Mora, F. Graziani, P. Teofilatto, C. Circi, M. Hechler, H. Yamakawa, J. Kawaguchi.

You will learn about other examples of the use of chaotic unstable motions to find low energy paths in later sections.

Chapter Eleven

Salvage of *HGS-1*, and a Christmas Present

. . . .

In early 1996 I was talking to a friend of mine, Howard Marks, in Westport, Connecticut, about the new route to the Moon and other interesting trajectories I had been studying for applications—not just in aerospace, but also in astronomy. Marks is a well-known entrepreneur in New York, and he is familiar with patents.

He asked me if I'd thought about patenting the route to the Moon used by *Hiten*. My initial reaction was to say I had, but that I'd assumed it was not possible because, first, I was working at JPL when I discovered it, and therefore whatever I did belonged to JPL. And second, the description of this route had been published in 1990, and according to patent rules, a patent has to be applied for within one year of its public announcement. Since it was 1996, and many publications had appeared on this, it was way beyond the deadline.

However, to my surprise, Marks said that it still might be possible using a type of patent called an algorithm patent. This is based upon a reproducible algorithm, which could be a computer algorithm.

An algorithm is an explicit procedure that can be broken down into a number of explicit steps that produce a specific result. It is reproducible if it is self-contained and can be independently verified. A computer algorithm is written in a particular computer language.

The reason he felt it might be possible to apply for a patent is that when the exterior WSB lunar transfer was computed for *Hiten*, it was based upon a backward algorithm. This is the backward time method we described in chapter 9 for determining *Hiten*'s transfer, a method that can be reduced to a set of computer commands comprising an algorithm. Marks knew that in early 1996, I had developed a new type of algorithm to determine the exterior WSB transfer that worked totally in forward time, starting from a desired point near the Earth, and going all the way to a desired point near the Moon. We call this a forward time method, or a forward algorithm.

He thought that I could apply for a patent based upon a forward algorithm. Marks knew a prominent up-and-coming patent attorney in Washington, D.C., Irah Donner, who would certainly know if it was possible to apply for a patent. He was just starting his career in patent law, and had already published an authoritative book on the subject. I spoke to Donner in January 1997, and after some investigation, he concluded it was indeed fine to apply for a patent on the lunar route under the framework of a forward algorithm. It was applied for a few months later along with several others based upon related ideas. In May 1997, under the guidance of Marks and the prominent law firm in Washington, D.C., Arnold and Porter, I formed a company called Innovative Orbital Design, Inc. (IOD, Inc.) to market the technologies associated with these patents. It turned out that the initial patent on the route to the Moon was granted in 2003. It represents the first patent on a route to the Moon [13]. By mid-2006, fourteen patents were granted both in the United States and internationally, with several still pending. These patents are now more generally stated, and have a wide application to a number of routes in space. One of these routes is discussed in the next section.

After forming IOD, Inc., in May 1997, I went to New York City and was introduced to prominent business investors by Marks, to try and get them interested in my company. The particular patent they were interested in involves a way of changing the inclination of an Earth-orbiting satellite using the Moon.

It works as follows: Suppose you have a satellite in Earth orbit and you want to change its inclination (i.e., the angle of its orbital plane with respect to the equator). A satellite changes its inclination by firing its rockets. This can require a lot of fuel, and it may not be possible if the rocket is low on fuel. However, there is a way to change the inclination indirectly, using less fuel, with certain restrictions. You first fire the rockets enabling the

satellite to leave Earth orbit and go to the Moon's weak
stability boundary, using an interior or exterior WSB
transfer. Now the weak stability boundary is a location
where the motion of a satellite is very sensitive. So, by firing the rocket engines for just a moment using a tiny
amount of fuel, you can cause a large change in the motion, so that it can come back to the Earth on a reverse
transfer, and arrive into Earth orbit with a completely different inclination. The fuel required to leave Earth orbit,
do a tiny change in the weak stability boundary at the
Moon, and be captured into Earth orbit again at the desired inclination can be relatively low. It can be less than
doing the inclination change while staying in Earth orbit
using standard methods. Using the exterior WSB transfer
is the desired transfer since it is the most flexible.

Potential investors in New York became interested
in seeing if it made sense to try and market to aerospace
companies this novel method of changing inclinations.
Companies with commercial Earth-orbiting satellites, such
as communications satellites for cell phones, television,
and so forth, looked the most interesting. The investors
were skeptical that this would be successful since it
seemed a bit risky and unusual. However, from their perspective, if it could be demonstrated to potential cus-
tomers that sufficient money could be saved, it might be
marketable as a licensing technology. One of the potential
investors, Henry Schachar of Centennial International,
LLC, in New York City, worked with me to try and interest people. By the end of 1997 Schachar was beginning to

feel that perhaps it was too much of a long shot. I was losing hope.

However, something happened to turn this around. On Christmas day 1997, a Proton rocket was launched from the Baikonur Cosmodrome in the Republic of Kazakhstan. It was carrying a large television satellite called *AsiaSat3* that was to provide service for India. The upper stage of the rocket malfunctioned and it didn't burn its rockets long enough and cut off early. They were supposed to burn for 110 seconds, and instead burned for only 1 second. So, instead of the satellite being placed in the desired orbit over the equator where it would remain fixed over India in a circular geostationary orbit (with an inclination of zero degrees), it was placed in a nondesired highly elliptical orbit with a large inclination of 52 degrees. So, a situation had occurred in which a fully functional satellite costing approximately $250 million was not in the desired operational orbit.

The possibility that *AsiaSat3*'s inclination could be changed using the Moon was was brought to my attention about a week later by Schachar who found out about it from Steven Gelles of Saga Investors, LP. I immediately did a calculation, and from the characteristics of the satellite saw that it did not have enough fuel by a large margin to change the inclination to zero degrees and circularize the orbit. However, there was enough to take it to the Moon on an exterior WSB transfer and then back to the Earth and into the desired geostationary orbit. I contacted a colleague of mine in California, Rex Ridenoure, who worked for the company Microcosm. Ridenoure knew all about the design of *AsiaSat3* since he had worked previously for the Hughes corporation that built it. He verified my calculations, and then contacted people at Hughes to see if they might be interested in helping IOD salvage *AsiaSat3* into the correct orbit using the lunar method. This would have been a perfect opportunity to get IOD off the ground.

Hughes immediately became interested in the idea, and they were able to purchase the failed satellite from the insurers who were stuck with it after it failed. I spoke to Hughes' orbital expert Cesar Ocampo in March 1998. He was to work with me as part of a cooperative agreement between Hughes and IOD that was verbally agreed upon. Shortly thereafter Hughes used a modified transfer to the Moon, more like a Hohmann transfer, but that still used aspects of the weak stability boundary. They successfully got *HGS-1* close to the desired geostationary orbit around the Earth. This was the first time this type of inclination change was done. However, Hughes failed to honor their cooperative agreement with IOD, or give acknowledgment to the origination of the idea—and it was disappointing to see a corporation of that magnitude behave in such a manner [44]. Fortunately, in 1998 two of the key technical experts at Hughes, Cesar Ocampo and Jeremiah Salvatore, together with myself and Rex Ridenoure shared the well-known Laurel award for this sponsored by *Aviation Week Space and Technology* magazine.

Chapter Twelve
Other Space Missions and Low Energy Transfers

. . . .

LGAS Reincarnated: SMART 1

The *LGAS* mission study at JPL was very successful, leading to a number of innovations in spacecraft and trajectory design. The spacecraft was built in an innovative modular design, and the components were miniaturized. Attempts to get NASA interested in actually making the LGAS mission into a real project were not successful primarily due to the long transfer time of about 1.5 years to go from the Earth to the Moon. A typical response was, "We've been to the Moon many times, taking only a few days to get there, so why should we go back and take almost two years?"

However, the European Space Agency (ESA) became interested in it in the early 1990s. I was consulting for them on low energy transfers at the large Italian aerospace company Alenia Spazio in Turino, Italy, and with ESA, headquartered in Darmstadt, Germany. They knew the potential importance of the exterior WSB transfer *Hiten* used, and the interior WSB transfer for *LGAS*. This led, in part, to the orbital expertise required for a lunar mission modeled largely on the *LGAS* design called *SMART 1* (Small Missions for Advanced Research in Technology) mission. *SMART 1* is ESA's first lunar mission and largely a technology demonstrator. Like *LGAS*, it will test new miniaturization technologies, and also uses solar electric xenon ion propulsion. ESA is interested in using ion engines in their future missions.

The mass of *SMART 1* is about 300 kilograms, which is similar to that of *LGAS's* which was about 200 kilograms. Its primary science mission is the same as *LGAS's*—namely, to look for water ice in shadowed craters at the Moon's south polar region by remote sensing. As with *LGAS*, *SMART 1* is designed to be low cost and serve as a basis for designing larger more ambitious planetary missions. The trajectory design is close to that of *LGAS*, where it is released from low Earth orbit and, using its small ion engines, gradually spirals away from the Earth and into ballistic capture at the Moon's weak stability boundary. It then spirals down into the desired lunar orbit taking a total time of about approximately 1.5 years.

SMART 1 was successfully launched on September 27, 2003, on the Ariane rocket of ESA from Kourou, French Guiana. It successfully arrived at the Moon into lunar orbit at the WSB on November 11, 2004. It was purposely crashed onto the Moon on September 3, 2006,

Figure 12.1 The SMART 1 spacecraft

ending its mission. The SMART 1 spacecraft is shown in figure 12.1.

The success of SMART 1 is very gratifying to me, as it represents a demonstration of LGAS. Even though my work on LGAS in 1986 eventually led to my departure from JPL, the arrival of SMART 1 at the Moon demonstrated that in spite of the long flight time, it was an interesting mission concept.

Europa Orbiter and Prometheus

Until late 1996 I had not thought about applying the notion of ballistic capture to any mission besides that of going to the Moon. The Moon had turned out to be a very difficult problem, and applying this method to another planetary capture problem would, I thought, be too challenging and time consuming. However, I had a welcome surprise at the annual Space Flights Mechanics Meeting in February 1997.

In the months prior to this time, the *Galileo* spacecraft orbiting Jupiter was making important discoveries. One of the most significant suggested that the Galilean moon Europa has an ocean of water just below the surface, maybe a few miles or less. In fact, the surface of Europa appears to be water ice, with cracks and a topography similar to what we see here on Earth near the north pole. The implications for this are enormous. The ocean could be warm due to the internal heating of the moon, and it would be protected from the powerful radiation around Jupiter. This means it is possible that life could have formed and evolved under the surface. Considering the unusual life that has evolved in hostile environments at the bottom of the Earth's oceans with huge pressures— near undersea volcanic vents, spewing lava and poison gasses—the idea of life forming within Europa seems not so far-fetched. This intriguing question naturally begs for a closer look at Europa, to see what is below the ice. To do that, you would have to send a spacecraft there to take direct measurements, carefully view the surface with high-resolution cameras, and perhaps even land and try to drill into the ice itself. I saw one NASA proposal that described a mission concept to explore below the surface of Europa in a small submarine.

Because of the interest in a mission to Europa, I was eager to hear a talk on a proposed mission by JPL to orbit Europa, aptly called *Europa Orbiter Mission* (EOM) by JPL mathematician and engineer Ted Sweetser. The proposed mission would send a spacecraft to Jupiter on the same type of trajectory that *Galileo* took. EOM would be sent into space as a payload in the space shuttle and released from the shuttle attached to a rocket stage, called a Centaur upper stage. This upper stage would be used to propel the spacecraft on its route to Jupiter, and the route would be indirect and rely on flying by the Earth and Venus to gain sufficient velocity to make it to Jupiter (see figure 12.2). This was necessary because of the limited capabilities of the Centaur stage. The Centaur stage would give enough velocity for the Europa spacecraft to escape the Earth's gravitational field and fall inward toward the Sun and fly by Venus. The gravity of Venus would sling it outward back toward and by the Earth, where it would gain more velocity. It then would travel out beyond the Earth, and only a small part of the way to Jupiter, loop back toward the Earth, fly by again, and to finally have sufficient velocity to make it to Jupiter. Once the spacecraft arrived at Jupiter, it would become captured by firing its rocket engines and go into a large elliptic orbit around Jupiter, far out beyond the orbit of Europa. Europa is one of the Galilean moons of Jupiter, which are, in order of increasing distance from Jupiter, Io, Europa, Ganymede, and Callisto. The initial capture orbit of the Europa spacecraft is shown in figure 12.3. The initial capture into orbit around

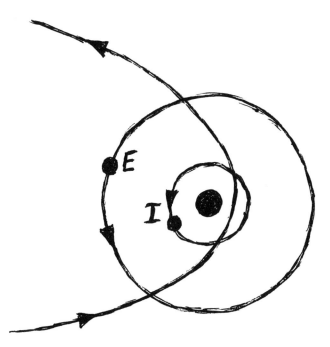

Figure 12.3 Initial capture of the *Europa* orbiter spacecraft into Jupiter orbit and the Galilean moons

Design for a Europa Orbiter: A Plethora of Astrodynamic Challanges." The next part of his lecture caught my attention, since he explained that capture into orbit around Europa was achieved by the same mechanism used by the lunar WSB transfers. The weak stability boundary between Europa and Jupiter is first estimated. Then, using the backward time method from the weak stability boundary, just as with *Hiten*, a path is found for the approaching spacecraft toward the weak stability boundary of Europa, and into a capture—requiring relatively low fuel. This was a brilliant application of the methods developed for the Moon to Europa. In fact, without the weak capture at Europa on a ballistic capture transfer, the Europa spacecraft would not have sufficient fuel to achieve orbit around Europa by standard means. This WSB Europa transfer is shown in figure 12.4.

The proposed Europa orbiter mission has evolved into a more ambitious mission concept, not just to orbit Europa, but the other Galilean moons as well, in a mission called *Prometheus* being developed at JPL. Its future is in doubt due to the high price tag. The *Prometheus* spacecraft does not use standard chemical rockets that were envisioned for the *Europa Orbiter Mission*, but instead has nuclear electric engines that use a nuclear generator to make the required electricity to generate a hot plasma for thrust. The development of these engines would be costly, and would drive up the total mission costs—to over a billion dollars. Such engines generate far more power than the standard chemical engines for a mission of this type, and enable more possibilities and flexibility. Solar power to generate electricity is not sufficient at Jupiter because the Sun is too far away, and so a nuclear generator must be used. Even if the *Prometheus* mission is canceled, it is likely that another more modest one will take its place, perhaps dedicated only to going to Europa, as was initially envisioned.

Exploration Initiative. In it he outlined a program to be taken, where we would first go back to the Moon and establish a lunar base, then go on to Mars with a manned mission. This is ambitious, but can be accomplished over an extended period of time if done in a well-planned manner.

This initiative has given rise to a set of studies on how to establish a lunar base, and for the construction of a special spacecraft called a Crew Exploration Vehicle (CEV), which would be used as a way to bring people to the surface of the Moon and beyond. I participated in one of these studies to design a low energy lunar transportation system for a CEV concept. Although the concept was not used, the system is interesting to describe.

The transportation system is used to get to and from the lunar base and Earth orbit with the CEV, which would use a Hohmann transfer to go to and from the Moon, since it would carry people and therefore require short flight times. However, it turns out that the exterior WSB transfer plays a surprisingly critical role. The CEV requires a lot of fuel to leave Earth orbit and go into lunar orbit. It cannot descend to the lunar surface, and leave the lunar surface to return to the Earth, unless it is refueled when it first arrives in lunar orbit. To refuel it, another spacecraft carrying fuel, called a Tanker Craft (TC), is required. The TC needs to arrive in lunar orbit using the least amount of fuel possible so as to have a sufficient amount of fuel to give the CEV. Also, the TC itself must have enough fuel to return to Earth. This sets up a tight constraint on how to design a path for the TC.

Figure 12.5 Lunar Base

It turns out that the use of the exterior WSB transfer by the TC solves the problem. The fact that it takes ninety days to get to the Moon is not an issue since it is only carrying fuel. The use of ballistic capture at the Moon saves the required amount of fuel necessary both to refuel the CEV and to return to Earth. This is an application of the use of the exterior WSB transfer that may play a key role in the establishment of a lunar base (see figure 12.5).

To return to the Moon and establish a base requires a new type of spacecraft. The space shuttle of the United States is designed to go into Earth orbit only. The CEV is currently being designed as a successor to the shuttle in 2010. It is called *Orion*. As the spacecraft to serve the nation, it will likely be expensive to operate, but having low-cost access to space is desirable, and until recently, such a goal seemed out of reach. One of the people in the study group I was involved in was Burt Rutan. He made history in September 2004 by building his own spacecraft, called *SpaceShipOne*, which a pilot used to fly into space on a suborbital path. The development cost was economical, a tiny fraction of what a government effort would be. This was the first time since the beginning of the space program that any private citizen had constructed a spacecraft to actually fly into space, and the feat is likely to open the door for public access to space. Someday this will open the door for seeing the golden arches of McDonald's on the Moon.

Chapter Thirteen

Hopping Comets and Earth Collision

. . . .

In 1987 I was doing a computer simulation of the motion of a tiny object that was ballistically captured by the Moon in the weak stability boundary. Since ballistic capture is generally temporary in nature, the object (say a rock), will move chaotically for a short of period of time, and then escape the Moon. I was curious about where it would go. My prior work with ballistic capture was focused on applications to spacecraft, wherein a tiny maneuver to stabilize its motion about the Moon can keep it captured for long periods of time. But suppose no maneuver were done? Since the WSB is a location where motion is chaotically unstable, I had no idea where the object would go, and really didn't expect that anything would be surprising. I figured the rock would rapidly escape from the Moon and go into some kind of typical elliptic orbit the Earth.

What I observed, however, was unusual. When I did

the computer simulation, it was easier for me to look at the printed numbers on the output sheet generated by the computer indicating where the spacecraft was with respect to the Earth and Moon than to make a graphical plot of the trajectory. There was much more information in the actual numbers, and from them I could visualize what the trajectory would look like. When I looked at the output sheet after the rock moved from ballistic capture, I could make no sense of the numbers. They seemed totally erratic and didn't look like any kind of typical motion around the Moon I had ever seen. I expected the object to perform a predictable elliptic-type motion around the Moon for a short time, while it was captured. Instead, there seemed to be no pattern to the motion at all. It moved in this way for about twenty days. Then, suddenly the numbers settled down indicating the rock had escaped the Moon and gone into an elliptic orbit around the Earth. This is illustrated in a plot of the distance of the rock to the Moon in figure 13.1. Was there an error?

Scratching my head, and feeling flustered, I decided to plot the trajectory. What I saw was completely unexpected. In backward time from the WSB starting at a point, A, at about 62,000 miles from the Moon, the rock went into an elliptical-looking orbit around the Earth. At first this elliptical orbit looked typical, but upon closer examination it wasn't. It is called a resonance orbit, an elliptical orbit that has the property that its period (that is, the time it takes the spacecraft to go around the Earth once)

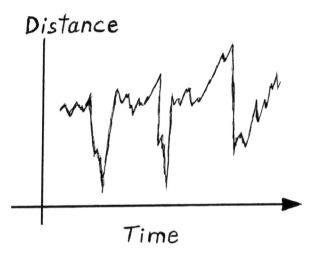

Figure 13.1 Erratic variation of distance around the Moon for a rock starting in the WSB

is synchronized with the period of the Moon, which is the time it takes for the Moon to go once around the Earth. In this case it was in a 2:1 resonance. The first digit is the number of times the spacecraft did a complete revolution on the ellipse, and the second digit is the number of complete times the Moon went around the Earth. So, a 2:1 resonance means that the spacecraft left its initial position A relative to the moon in the WSB, and went around the Earth twice on this ellipse. Meanwhile, the Moon went around the Earth once. So, at the end of the second cycle of the spacecraft it was back very near to where it had started, approximately at A with respect to the Moon. Now, this was just described in backward time. In forward time from the WSB start at A, it will therefore also be in the same 2:1 resonance orbit and end up at the initial WSB start at point A again. This is shown in figure 13.2. This 2:1 orbit lies completely interior to the orbit of the Moon around the Earth.

Now, what happens in forward time from the initial WSB start at point A? This is where it gets really interesting. The spacecraft was abruptly pulled by the Moon's gravity into a very complicated oscillating motion around the Moon. This oscillating motion was out of the initial planar motion it had in the 2:1 resonance orbit, which was approximately in the plane of the Moon's orbit around the Earth (i.e., the Earth-Moon plane) and actually perpendicular to it. It was chaotically moving with respect to the Moon in the WSB where it felt a near equal pull on it by the Earth. This motion is shown in figure 13.3. It performs this unusual motion for thirty-four days, corresponding to a little more than one lunar period of twenty-eight days. It is then abruptly ejected away from the Moon into another much larger elliptical orbit, orbiting the Earth also in the Earth-Moon plane, lying now completely outside the orbit of the Moon around the Earth. A closer examination showed that this was also a resonance orbit, of type 4:5. So in the time the spacecraft went around the Earth four times, the Moon went around the Earth five times. The full motion described is shown in figure 13.4.

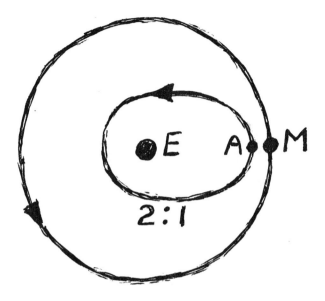

Figure 13.2 A 2:1 resonance elliptical orbit around the Earth
in resonance with the Moon

In summary, the following motion was then observed,
starting from point A in the WSB:

2:1 resonance → chaotic oscillation in the WSB → 4:5
resonance

I had never seen this kind of motion before. In this
simulation, only the gravitational fields of the Earth and
Moon were modeled. This kind of changing of resonances

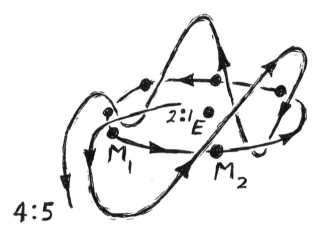

Figure 13.3 Chaotic motion in the WSB of the Moon during a resonance hop transition

is called a resonance transition or resonance hop from one resonance to another.

From my experience with theoretical celestial mechanics and dynamical systems, I knew that this was an unusual and interesting motion that had heretofore not been observed. It was contrary to how resonance transitions were known to occur. The known process that was proved to exist in 1964 by the Russian V. Arnold is called Arnold diffusion. In that motion, a spacecraft can transition between resonances, but the time scale to do so should be on the order of many millions of years—not thirty days! The process that I witnessed is called a resonance hop to emphasize the extremely fast speed. From when it

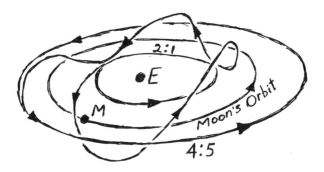

Figure 13.4 A resonance hop from a 2:1 resonance to the lunar WSB to a 4:5 resonance

was discovered in 1988 until 1995 I had shown it to a number of mathematicians, and there was a good deal of interest. In particular, Princeton mathematician John Mather and French mathematician Michel Herman showed some interest in this when I showed it to them in 1994 and again in 1996. I remember Herman asking me in 1994 if there was possibly an error in my modeling.

A major insight in understanding this process came in April 1995 when I was invited to give a lecture at a conference in New York at the United Nations on the subject of near Earth objects. At the very end of my lecture, I presented the hop process just described, and after I presented it, I said that although it is very interesting, it may not be correct or physically of interest. As soon as I had said this, the Harvard astronomer Brian Marsden enthusiastically pointed out that comets have been observed to perform this hopping motion. Instead of a spacecraft performing a hop by interacting with the lunar WSB, we now have a comet performing a hop by interacting with the Jupiter WSB. In this case, the bodies under consideration are the Sun, as the major large body, Jupiter as the smaller planetary body, and a comet, which has negligible mass with respect to the Sun and Jupiter. This is analogous to what was previously considered when the Earth was the larger body, the Moon the smaller body, and the particle of negligible mass a rock.[1]

This connection by Marsden was a concrete example of the hop motion, and it led to a collaboration on the investigation of comets that perform hop motion. We tried to tie it together with my work on the WSB. In 1997 we published our results, "Hopping Motion in Comets," in the *Astronomical Journal*.

Let's elaborate on the hopping comets. Jupiter's WSB, with respect to the Sun, extends out about 37 million miles! A comet moving in the WSB of Jupiter feels the gravitational pulls on it of the Sun and Jupiter almost equally, so its motion is chaotically unstable.

If a comet is moving in an ellipse around the Sun, and that ellipse passes close enough to Jupiter and with the right velocity, it can pass within the WSB of Jupiter.

---

[1] Jupiter is one thousandth the mass of the Sun! Jupiter's orbital period of motion around the Sun is 11.86 years. A comet, typically the size of Manhattan island, has negligible mass with respect to the Earth, and the Earth is one thousandth of the mass of Jupiter.

When this happens, the comet can be abruptly pulled into an unstable orbit around Jupiter, where it is weakly captured. Soon after being captured, it will generally escape Jupiter and pass onto another elliptic orbit around the Sun (since the comet escapes Jupiter, this capture is also called a temporary capture). In this case the elliptic orbits around the Sun that the comet had before and after the interaction with Jupiter are in resonance with Jupiter. Comets therefore perform a hop from one resonance to another and they belong to the class of short period comets where their period of motion around the Sun is less than 200 years. The resonance hopping comets have periods that lie between 7 and 52 years. This contrasts with comets of periods of thousands of years, which travel many times farther from the Sun than Jupiter. The durations of the hops, that is, the WSB interactions with Jupiter, have been observed to vary between 17 days and 17 years. A good example of a resonance hopping comet is Gehrels 3, shown in figure 13.5. It is shown starting in a 2:3 resonance ellipse around the Sun with a period of 18.2 years moving exterior to the orbit of Jupiter. On November 6, 1968, it started a hop that lasted until December 5, 1973, when it transitioned into a 3:2 resonance ellipse of period 8.1 years.

The complexity of a comet motion during a hop can be observed in the comet Helin-Roman-Crockett, which we see in figure 13.6. This comet started in a 3:2 resonance elliptical orbit around the Sun, performed a hop, and transitioned onto another 3:2 resonance orbit. The

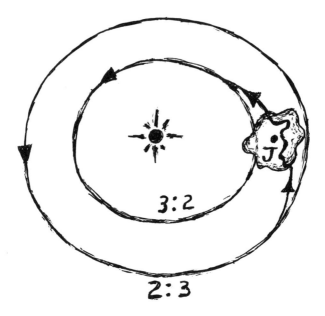

Figure 13.5 Resonance hop of the comet Gehrels 3, transitioning from the outer 2:3 resonance orbit to the inner 3:2 resonance orbit around the Sun

hop started on December 5, 1968, and lasted until August 3, 1984.

A resonance transition with Jupiter for hopping comets has never actually been observed, in part because it is very difficult to see a comet when it is Jupiter's distance from the Sun. Comets tend to be dark and are most observable when they are nearer to the Sun and have their characteristic glowing coma with a tail. They can be

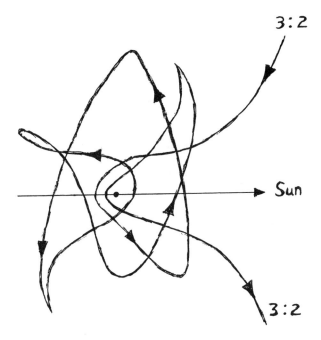

Figure 13.6 Resonance transition of the comet Helin-Roman-Crockett, transitioning from a 3:2 resonance orbit to another 3:2 resonance orbit around the Sun. This figure is Jupiter centered where the Sun is fixed on the positive horizontal axis

observed at Jupiter when they are suitably lit by the Sun. For example, Oterma was observed on March 27, 1943, at the Turku Observatory in Finland by Liisi Oterma. It was already moving in its 3:2 resonant orbit around the Sun with a period of 7.9 years. It was only through doing a nu-

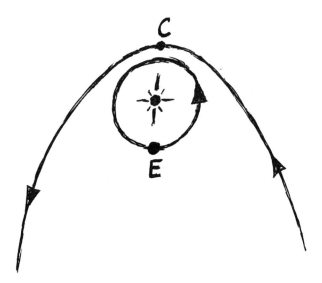

Figure 13.7 Comet at conjunction with the Sun

merical simulation with the computer in backward time
that it was discovered it had actually passed near Jupiter
on November 23, 1936, from a 2:3 resonant orbit of period
18.2 years, moving outside Jupiter's orbit around the Sun.
It then performed a hop, moving near Jupiter until December 26, 1938, and then it transitioned onto a 3:2 resonant
orbit, interior to Jupiter's orbit around the Sun. The same
numerical simulations predicted it would perform another
hop on May 21, 1962, lasting till February 12, 1964, when
it would transition back to the 2:3 elliptical orbit of period
18.4 years. It was not possible to observe this because when

TABLE 13.1

Resonant hopping comets and resonance transitions

Comet	Resonance transition
Helin-Roman-Crockett	3:2 → 3:2
Harrington-Abell	5:3 → 8:5
Oterma	2:3 → 3:2
Gehrels 3	2:3 → 3:2
Lexell	5:4 → 2:1
Smirnova-Chernykh	6:13 → 7:5
Kearne-Kwee	3:13 → 4:3
Pons-Winnecke	2:1 → 2:1
Wolf	7:4 → 3:2 → 4:3 → 3:2 → · · · → 4:3

this transition occurred, Oterma was in conjunction with the Sun, that is, the Sun was between Jupiter and the Earth, blocking the view (see figure 13.7).

The comets listed in this section (table 13.1) have had sufficient observational data to construct reasonably accurate trajectories for a relatively short period of time, say about fifty years. It is difficult to obtain reliable trajectory data from numerical simulations of these comets for longer periods without additional observations. This is because when the comets are performing a resonance hop, they are moving chaotically, and extremely small variations in the motion due to subtle perturbations not mod-

eled can cause large deviations from the true trajectory on
the computer.

Potential Earth Collision

We know that a comet on a resonant orbit can collide with
a planet. This was the case with the comet Shoemaker-
Levy 9. It made a spectacular collision with Jupiter in July
1994, and as it approached Jupiter it broke up into many
pieces. Some of the explosive impacts with Jupiter created
explosions the size of the Earth! This comet had a type 2:3
resonant orbit similar to Oterma's when the former was
weakly captured by Jupiter before collision, before it had a
chance to perform a hop to another resonant orbit, as, for
example, Oterma did. The collision of Shoemaker-Levy 9
with Jupiter demonstrates that the Earth may be at risk of
collision with a comet while the latter hops from one resonance orbit to another.

Can a resonant hopping comet collide with the Earth?
If this were to occur it would destroy most life, in the same
way it is conjectured a comet or asteroid impact did approximately 60 million years ago, leading to the extinction of
the dinosaurs. Such an impact can unleash the equivalent
of billions of hydrogen bombs! So it is important for our
own survival on this planet to understand this question.

It is indeed possible for a resonant hopping comet to
collide with the Earth, and one came close in the 1700s,
the comet Lexell, which is listed in table 13.1.

Lexell

In 1770 Charles Messier observed an object that turned out to be a comet, and using Messier's observational data, Johann Anders Lexell computed the orbit and determined that the comet was in an elliptical orbit around the Sun. He also determined that prior to being in elliptical orbit, and before its discovery, the comet's trajectory was altered by Jupiter's gravity in 1767. He predicted that it would be gravitationally perturbed by Jupiter again in 1779, in a more dramatic way. This comet was named after Lexell.

Although the comet was scarcely visible in a telescope on June 14, 1770, when Messier observed it, he was able to see it with the naked eye on June 22. It was observed throughout the world, and could even be seen in daylight. On July 1 the comet passed its closest distance to the Earth, of only .0146 AU (1 AU = 1 astronomical unit = distance of the Earth to the Sun = 93,000,000 miles). This is only 1,357,800 miles, which in astronomical terms is a near miss. As mentioned, if a comet of this size hit the Earth, it would cause serious problems for all life on the planet.

Later numerical simulations of the trajectory of Lexell confirmed the earlier calculations of its motion [8] and show the following: In 1722 Lexell was in an approximate resonant 5:4 orbit around the Sun with a period of 9.3 years. This orbit was fairly eccentric—it looked like a thin ellipse. It had a periapsis distance with respect to the Sun of 2.9 AU

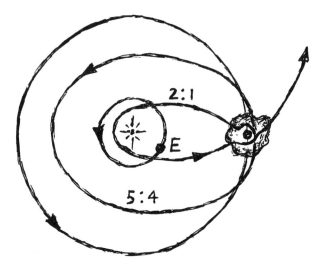

Figure 13.8 Trajectory of Lexell from 1722. It is initially in the
larger 5:4 elliptical orbit of period 9.6 years. In 1755 it hops
onto a 2:1 Earth orbit crossing ellipse, flying by the Earth at
only 1.36 million miles

and an apoapsis distance of 5.9 AU, slightly beyond the orbit of Jupiter, which is about 5.2 AU. This elliptical orbit is
shown in figure 13.8.[2] On January 12, 1767, Lexell started a
hop near Jupiter that lasted until June 9, 1767. It transitioned onto an approximate 2:1 resonance ellipse with a period of 5.6 years. This ellipse has a periapsis distance from
the Sun of only .67 AU and crosses the orbit of the Earth.

―――――――――

[2] The periapsis of an elliptical orbit around the Sun is the
closest point on the orbit to the Sun. The apoapsis is the farthest
point to the Sun.

Thus, Lexell was transformed into an Earth orbit-crossing comet by a resonance transition. This transformation took only five months, and to an observer on the Earth this comet would appear to have suddenly come from nowhere. While on the 2:1 ellipse, it approached the Earth and came within 1.36 million miles (only about five times the distance of the Moon to the Earth!) on July 1, 1770. After that, Lexell moved out to a close approach with Jupiter, on July 27, 1779. It flew by Jupiter at a distance of only 135,780 miles, which significantly increased the velocity of the comet through a gravity assist using Jupiter's powerful gravitational field. This flung the comet onto an outward path from the Sun, which was on a huge ellipse with a period of 337 years and an apoapsis of 91.5 AU as illustrated in figure 13.9. This is far beyond the orbit of Pluto (Pluto is approximately 40 AU from the Sun) in the region called the Kuiper belt, where many planetoid-like objects, of which Pluto is thought to be one, orbit the Sun.

Jupiter-Hopping Earth-Crossing Comets Present a Danger

The trajectory of Lexell needs to be better understood, as Lexell represents a class of comets that can abruptly be changed into possible Earth-colliding comets by a hop, and with very little warning. The hop lasted only five months and it flew close to the Earth only three years

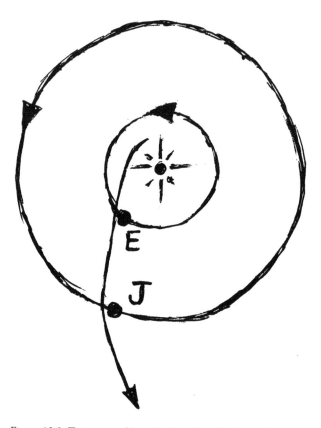

Figure 13.9 Trajectory of Lexell after it flew by the Earth in
1770, where it goes out to fly by Jupiter and then out beyond
Pluto to the Kuiper belt

later. Three years is all the warning we would have (in fact, we would have no warning at all if the comet wasn't observed until it was close to the Earth). Since this comet ended up going into the Kuiper belt, it is possible that cometary objects could originate in the Kuiper belt, pass near Jupiter, and become Earth crossers with little warning. This implies that it might be wise to do a survey and try to understand the population of such objects, which at this time is not known.

Kuiper Belt Objects and Neptune Hopping

The process of comets changing resonance types via WSB interaction with Jupiter can be applied to Kuiper belt objects passing through the WSB of Neptune. A large number of objects have been observed traveling out beyond the orbit of Neptune between 30 and 50 AU since the first one, called 1992 QB1, was discovered in 1992 by Harvard astronomers D. Jewitt and J. Luu. The second one was discovered in 1993 by A. Fitzsimmons, I. Williams, and D. O'Ceallaigh, and called 1993 SC. The belt is named after the Dutch astronomer Gerald Kuiper, and the objects in it are not well understood, since they are so far away. Some can be quite large, as for example 2002 LM60, referred to as Quaoar, which has a diameter of about 806 miles, a little over half the size of Pluto and larger than Pluto's moon Charon. It is hypothesized that Pluto—a relatively small object, only about 18 percent of the Earth's diameter, and about 60 percent of the diameter of our own Moon—is one of these objects.

A substantial percentage of Kuiper belt objects move in elliptical orbits that are in resonance with Neptune, as is the case with Pluto, which moves in an approximate 2:3 resonance with Neptune. Pluto, whose distance from the Sun of about 40 AU, also lies in the zone for Kuiper belt objects. There is evidence that Kuiper belt objects in 2:3 resonance with Neptune may have encountered Neptune's WSB in the distant past and performed a hop into their current 2:3 orbits. This can be shown as possibly the case by assuming that at some point in the distant past a Kuiper belt object was weakly captured by Neptune—in its WSB—and hopped from one resonance into its current one of 2:3 resonance. When this is assumed, the eccentricities of the orbits can be predicted. The difference in the calculated eccentricities from their observed values averages .012. Since this is such a small difference, it supports the premise that these objects may be Neptune hoppers. There are possible implications for us, since a Kuiper belt object could encounter Neptune's WSB and perform a dramatic resonance transition, which could put it on an elliptical trajectory around the Sun that could cross the Earth's orbit and collide with the Earth. This process would be analogous to comets interacting with Jupiter's WSB and becoming Earth orbit-crossers like Lexell. Kuiper belt objects can be quite large, and at least one we know, discussed below, is even bigger than Pluto. If a large planetary-sized object collided with the Earth, the outcome would not just threaten all life on Earth, but would alter the Earth's orbit around the Sun. Although this scenario may be unlikely, it is theoretically possible. As with a comet, a collision between the Earth and a Kuiper belt object would eliminate most life on the Earth.

Ballistic Escape from the Earth-Moon System, and Asteroid Capture

There is an interesting process related to resonance hopping that was noticed in 1987. Figure 13.4 shows a hop from a 2:1 elliptic motion to a 4:5 ellipse around the Earth in resonance with the Moon. This is modeled using the Earth and Moon as the primary large masses, and an object of insignificant mass relative to the Earth and Moon, that is, a spacecraft, performing the resonance transition. So, the motion is modeled by a three-body problem between the Earth, the Moon, and the spacecraft. For motion like this in the Earth-Moon system, modeling the gravitational perturbation of the Sun has a substantial effect, since the motion of the spacecraft in the lunar WSB is so sensitive. The original 2:1 elliptical motion around the Earth remains stable when the Sun is factored in; however, the chaotic hop dynamics shown in figure 13.4 breaks down. The Sun pulls it apart, and instead of the spacecraft transitioning to the 4:5 elliptic motion, it is pulled out to 930,000 miles from the Earth, where it enters the Earth-Sun WSB. When it is in this region, the gravitational pulls

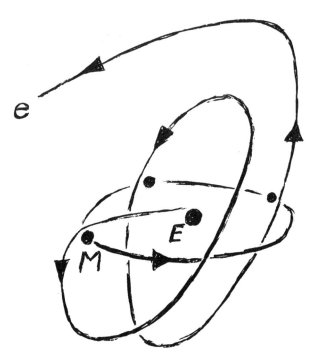

Figure 13.10 Ballistic escape from the Earth-Moon system

of the Earth and the Sun approximately balance, and the
Sun pulls the spacecraft away from the Earth-Moon system
and into orbit around the Sun! In this way, the spacecraft
is able to escape the Earth-Moon system with no spacecraft
rocket firings, which is usually required to escape the Earth-
Moon system. This is called ballistic escape, and is shown
in figure 13.10, labeled **e**. This observation is interesting

because it shows how to save substantial fuel to escape the Earth-Moon system!

When the spacecraft escapes the Earth-Moon in this manner, the process takes about 180 days. The escape process is also gradual, and when the spacecraft is in Earth-like orbit around the Sun, it has a velocity close to that of the Earth's around the Sun of about 17 miles per second. If one wished to transfer the object beyond Earth orbit, say to Mars or another planet, then this escape mechanism could be augmented by firing the rockets during the escape to gain additional speed. A modification of this approach was used by the Japanese Mars mission *Nozomi*, formally called *Planet B*. This was a mission to place *Nozomi* in orbit around Mars in 1999. After launch from the Earth in 1998, the spacecraft went to the Moon on a Hohmann transfer, flew by it, returned to fly around the Earth, and then flew by the Moon again. These lunar flybys were gravity assists that gave the spacecraft enough speed to fly out to the Earth WSB at 930,000 miles from the Earth. *Nozomi* then fell back close to the Earth and did an Earth flyby for another gravity assist to gain enough speed to go to Mars. However, due to a faulty fuel valve the spacecraft did not have sufficient fuel to directly go to Mars in 1999 or achieve orbit upon arrival. In order to make it to Mars, it had to go on a longer nondirect route, arriving in 2003, when it flew by at a distance of about 620 miles. It is now in a Mars-like orbit around the Sun.

Ballistic escape can be used as a way to reduce the costs of an eventual manned Mars mission. Supplies could be sent separately on this route, taking about half a year longer than a Hohmann transfer. This would save mass on the manned spacecraft, which would take the Hohmann transfer.

The reverse of this ballistic escape process could provide a way to capture an asteroid into orbit around the Earth if it was in an Earth-like orbit around the Sun. In this case, looking at the reverse of the motion shown in figure 13.10, it would be pulled into the Earth-Moon system starting at approximately 930,000 miles away, and then fall into several flybys of the Moon, and then settle into a 2:1 resonant orbit around the Earth. It might be possible to modify the orbits of some of the asteroids that cross Earth's orbit and perform this capture process, which would be desireable because an asteroid orbiting the Earth could be mined as a resource for valuable minerals that the Earth may someday need.

Chapter Fourteen

The Creation of the Moon by Another World

. . . .

The use of low energy trajectories and weak stability boundaries has another interesting application, and it is related to the origin of our Moon.

An outstanding question in astronomy has to do with understanding where the Moon actually came from. One of the first theories to try to answer this is the sister planet theory. It proposes that the Moon and the Earth were sister planets formed in the solar nebula of gas and dust from which all the planets formed about 4 billion years ago. However, there are some inconsistencies with this. One is that a large iron core is absent from the Moon and present in the Earth, giving the Earth and Moon different densities: 5.5 grams/cm^3, 3.3 grams/cm^3, respectively. Another theory is that the Moon was formed from beyond the Earth's orbit and was captured into orbit around the Earth. If this were the case then the Earth and Moon would have different abundances of oxygen isotopes.[1] This is inconsistent with the fact that the Earth and Moon have identical abundances.

A generally accepted theory that explains the differences in iron and the oxygen isotope abundances, among other things, is called the impactor theory. It was formulated by W. Hartmann, D. Davis, and A. Cameron, W. Ward. It proposes that after the Earth had formed 4 billion years ago, a giant Mars-sized object smashed into it, forming the Moon from iron-poor mantle material debris primarily from the impactor, and also from the Earth—both of which already had iron cores. This explains the iron deficiency of the Moon. This theory also proposes that the impactor formed at the same 1 AU distance that the Earth is from the Sun, which explains the identical oxygen isotope abundances in the Moon and Earth. This impact is sometimes referred to as the big splat, illustrated in figure 14.1. The impactor strikes the Earth with a relatively slow velocity on the order of a few hundred meters per second, which is called a near parabolic collision. Such a collision between two large planetary objects would be awesome to observe—from a safe distance!

A fundamental question to ask is, Where did the giant Mars-sized impactor come from? In May 2001 while I was attending a conference at Princeton University, Princeton astrophysicist Richard Gott described his hypothesis. He

---

[1] Isotopes are different versions of an element having the same number of protons, but a different numbers of neutrons.

Figure 14.1 A Mars-sized impactor smashing into the Earth. Painting by Dr. William K. Hartmann, Planetary Science Institute

proposed that at the time of the solar nebula from which the Earth was formed, there was so much debris flying around the Sun that it could have settled into the stable equilateral Lagrange points L_4, L_5 with respect to the Earth and Sun. These are shown in figure 14.2. Since these locations are stable, any debris arriving at them with a small relative velocity would remain trapped. As more and more debris arrived, it could have started to coalesce and a massive body could have begun growing. Given

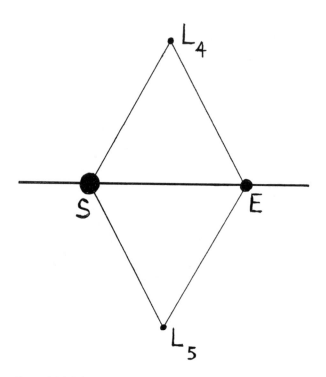

Figure 14.2 The equilateral Lagrange points in the Earth-Sun system

several million years, a large Mars-sized body could result. After Gott explained this to me, he posed the question of how this object, once formed, could find itself on a collision trajectory with the Earth. Obviously this piece of the puzzle is necessary, and the answer was not obvious. After mulling it over for several months, the solution became clear. Just as a planetary body, say the Earth, under the gravitational perturbation of a larger body, the Sun, has a weak stability boundary, L_4 (or L_5) should also have a weak stability boundary, which for convenience we will label as WSB_L. Remember that a weak stability boundary should be thought of in terms of both location and speed, as we described in chapter 7. So at a given distance, d, from L_4 (or L_5), out to a maximal distance (approximately 62,000 miles in the case of the Moon), there is associated a critical speed, or velocity, say $V(d)$, depending on the angular direction from L_4 (or L_5). For brevity, we'll focus only on L_4. If the velocity of a small body is less than $V(d)$, it will tend to be trapped around L_4, and for a velocity greater than $V(d)$, it will escape L_4. As we discussed in chapter 7, the WSB can be thought of as a collection of irregularly shaped surfaces around the Moon, one inside the other. As these surfaces get smaller in size the associated values of $V(d)$ get larger. If we are located exactly at L_4, we'll need the largest velocity $V(0)$, which, in this case, is $d = 0$. This is the critical escape velocity from L_4. With this velocity, depending on the direction you are leaving L_4 from, you will be at the transition between capture and escape from L_4 (see figure 14.3). This represents the WSB_L surface that prescribes the required velocity necessary, depending on the direction of motion from L_4, to be weakly captured.

Thus, if the velocity at L_4 is only very slightly greater than $V(0)$, the small body, say the impactor, will just barely escape L_4. We call this weak escape. The motion of

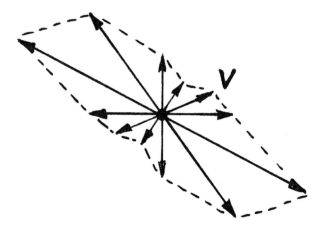

Figure 14.3 WSB$_L$ surface prescribing critical velocities depending on the direction of motion from L_4 (or L_5). The surface in this case is a curve, which we see is irregular

a small body leaving L_4 will be chaotic for the velocity approximately equal to V(0). An analogy would be filling a bathtub with water. As the water is slowly added to the tub, the tub starts to fill. When it gets near the top, and water is put in too fast, it will rapidly spill out; however, if water is added slowly, when it reaches the top of the tub it will just barely flow over the top. Now once the impactor just barely escapes L_4, it will move slowly away from L_4, and the motion will remain approximately at the same distance from the Sun as L_4; that is, 1 AU, which is the Earth's distance from the Sun. Eventually the impactor will reach the Earth either from the front or the back. Since the impactor lies about 1 AU from the Sun, it will pass very close to the Earth and either be gravitationally pulled in to collide with the Earth with a slow relative velocity, as desired, or will closely fly by getting a gravity assist from the Earth, and re-encountering the Earth. The motion of the small body in this orbit is very sensitive, and it seems that collision would be likely.

When this idea was implemented on the computer, it indeed produced trajectories that slowly escaped from L_4, crept slowly along Earth's orbit around the Sun, and collided with the Earth in a near parabolic collision consistent with the impactor theory. When the impactor is slowly moving along the Earth's orbit, its motion is sensitive, and so we call this motion chaotic creeping collision trajectories. This is shown in figure 14.4. The mechanism for the impactor to leave L_4 is the following: As debris collects at L_4, the growing object will have collisions or near collisions with the debris. These encounters will cause the growing impactor to move around near L_4, and it will gradually increase its velocity with respect to L_4 over time. After several million years it will simultaneously have the required size and velocity to put it on a collision course with the Earth taking on the order of one hundred years. This is described in detail in a research paper published by Belbruno and Gott entitled, "Where Did the Moon Come From?"

A hypothetical observer on the Earth 4 billion years ago would start to notice a bright object in the night sky 60 degrees in front of or behind the Earth in its orbit. Over time it would get brighter and brighter. Eventually it

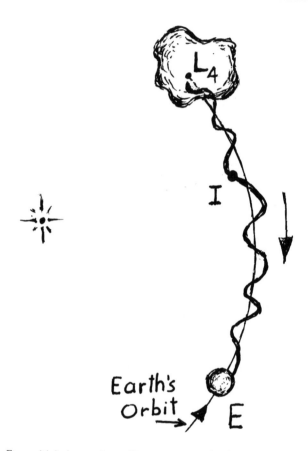

Figure 14.4 A creeping collision trajectory for the giant impactor, **I**, from **L₄** to Earth collision

would start to move and increase substantially in brightness. It would be seen in daylight as a bright object in the blue sky. It would increase in size until it appeared large in the sky. A relatively short time later, it would fill the sky and then fall in a cataclysmic collision, and the Moon would be born. Thus, the Moon can be thought of as having been created by another world.

This proposed theory for the origin of the impactor may have applications to the formation of moons of other planets. At this time, it is an open question, recently investigated by Richard Gott, Robert Vanderbei of Princeton University, and myself. Our results indicate that some of the moons of Saturn may have been formed at particular equilateral Lagrange points within the Saturnian system. For example, numerical simulations by Vanderbei show that the moons Janus and Epimetheus have the motions observed today if they initially started at these Lagrange points.

Chapter Fifteen
Beyond the Moon and to the Stars

. . . .

Pluto to Alpha Centauri

Throughout this book I have discussed low energy chaotic motions in different situations. These motions include ballistic capture and escape, resonance transitions, and creeping collision orbits. These have been applied to the motions of spacecraft, comets, asteroids, Kuiper belt objects, and planet-sized impactors, and have resulted in new ways to reach the Moon, explore the moons of Jupiter, rescue errant satellites and missions, escape the Earth-Moon system, propose a theory on the origin of the Moon, shed light on resonance hopping comets, and possible hops in the Kuiper belt. These applications to space missions and planetary astronomy only touch the very tip of an iceberg, as these dynamics and motions can be applied in many other situations as endless as there are planets and moons. The complexities of motion increase when we go from a two-body problem to a three-body problem, and then to a four-body problem. The latter two are not well understood and likely contain many more surprising motions.

One characteristic of the motions presented in this book is that although they are low energy in nature, they represent slow processes. For example, whereas a Hohmann transfer to the Moon is three days, the exterior WSB transfer takes three months. This is because these motions are occurring at the fringes of stability regions, where the motion is delicate and sensitive. As with a glider plane riding the wind currents, these motions ride the subtle tugs and pulls of gravitational fields between planetary bodies.

These methods are ideally suited to explore sensitive chaotic motions associated with comets, asteroids, planetoids, and other heavenly objects since they move within the natural gravitational fields. They are also ideally suited if we want to save fuel for spacecraft missions when we are not in a hurry to get to our destination, and to save fuel when we arrive. For example, if we decide to go to the Pluto-Charon system, we want to get there as quickly as possible since it is so far away. A Hohmann transfer could take about twelve years from the Earth. A low energy transfer slowly meandering among the various gravitational pulls of the planets would be prohibitively long, and could take hundreds or even thousands of years. However, upon arrival, the Pluto-Charon system would be ideal for low energy trajectories since Pluto and Charon are comparable in size, which implies that many chaotic motions exist within this system. So in this case, low energy motions, which use less fuel, could enable a spacecraft to explore this system for longer periods of time. However, in the case of going to the Moon, it isn't terrible to take three months to get there by WSB transfer instead of three days by a Hohmann transfer. It is also possible that future research may find WSB transfers with faster flight times.

Our solar system beyond Pluto is not very well understood. Kuiper belt objects exist from approximately 30 AU to 50 AU. It is hypothesized that the Kuiper belt is a source of short period comets, which we saw could be potential hoppers at Jupiter or Neptune. Many thousands of objects are thought to exist in the Kuiper belt with diameters ranging from a few miles to almost a thousand, such as Quaoar. In fact, a large Kuiper belt object, known as 2003UB313, was announced in July 2005 as a candidate for the tenth planet of our solar system. It is about 1.5 times larger than Pluto and inclined to the ecliptic (the Earth's orbit plane) by 45 degrees. This is unusual since the planets are generally in the ecliptic plane, an exception being Pluto, which is inclined by 17 degrees. The orbit of 2003UB313 at the time of its discovery was at a distance of 97 AU from the Sun, well beyond Pluto, which was at a distance of 31 AU. The planetary status of 2003UB313 depends on whether Pluto maintains its designation as a planet which seems unlikely. Like 2003UB313, Pluto is now known to be a Kuiper belt object, and the astronomy community has recently decided that both Pluto and 2003UB313 are classified as dwarf planets. Currently, the number of planets in our solar system is reduced to eight, together with a number of dwarf planets. It is likely that more objects the size of Pluto or larger will be discovered in these highly inclined orbits around the Sun. An interesting object, called Sedna, has a diameter estimated at between 620 and 930 miles, and is in a highly elliptical orbit with a periapsis to the Sun at 76 AU and an apoapsis of approximately 1,000 AU. This object lies outside of the Kuiper belt and represents a new class of objects between the Kuiper belt and the Oort cloud, the latter named after the Dutch astronomer who hypothesized its existence. The Oort cloud, thought to be the source of long period comets, is a region that extends roughly between 10,000 and 100,000 AU from the Sun, and is conjectured to contain trillions of comets. The distance of 100,000 AU is immense, yet it is only 4 percent of the distance to the nearest star system.

Beyond the Oort cloud region, we are exiting our solar system and entering interstellar space. The nearest star to the Sun is the triple star system of the stars Alpha Centauri A, Alpha Centauri B, and Proxima Centauri (collectively called Alpha Centauri) which is 4.3 light years away[1]. This is very far away by human standards. Since one light year is about 63,240 AU, which is 1,581 times

[1] A light year (LY) is the distance light travels in one year at a speed of about 183,000 miles per second (i.e., 300,000 kilometers per second) or, equivalently, seven times around the Earth in one second; 1 LY = 5,890,000,000,000 miles.

the distance of Pluto, then Alpha Centauri is 6,798 times the distance of Pluto. To get an idea of just how far this is, let's estimate how long it would take a spacecraft to reach Alpha Centauri with our current technology. Using a small robotic spacecraft and the most powerful rockets we have, and by flying by Jupiter to gain velocity through a gravity assist, we can reach Pluto in about ten years. So, to reach Alpha Centauri, it would take 67,980 years! Thus, reaching the stars by conventional rockets is currently not possible.

If we are to send spacecraft to the nearby stars in relatively short time spans, within a human lifetime, a radically new type of rocket engine must be designed that would herald in an unprecedented new age of exploration to new star systems and unimagined discoveries.

Comets Moving between the Sun and Alpha Centauri

Even though it is unrealistic to send a conventional spacecraft to Alpha Centauri, it may be possible that comets can move between the Oort cloud region of the Sun and a similar region around Alpha Centauri.[2] We need to assume that we have a comet going around the Sun in a highly elliptical orbit, with an apoapsis far from the Sun at 100,000 AU (in the Oort cloud). This comet is

[2] There is a way for this to happen, based on the work in the last chapter of my book *Capture Dynamics and Chaotic Motions in Celestial Mechanics*.

assumed to move in Jupiter's orbital plane around the Sun. We also need to assume that the comet flies by Jupiter just slightly faster than required for weak capture—so it flies by relatively slowly. Given these assumptions, the motion of the comet would be chaotic, which means in this situation that when it passes by Jupiter, moves around the Sun, then out to 100,000 AU, it need not remain a highly eccentric ellipse. When it is very far away, it can appear parabolic, approaching a zero velocity at 100,000 AU relative to the Sun and never return to orbit the Sun, or it can appear hyperbolic, approaching a positive velocity. In this case, the comet would move beyond 100,000 AU and out of the Oort cloud to escape the Sun's gravitational pull. It could theoretically escape with a velocity of a few miles per second, and head in the direction of Alpha Centauri. At a few miles per second, say for sake of argument, 1.24 miles (2 kilometers) per second, it would take the comet 645,000 years to arrive at the outer reaches of the Alpha Centauri system at approximately 100,000 AU from Alpha Centauri. This provides a mechanism for comets to pass from our solar system to the Alpha Centauri and is illustrated in figure 15.1.

When the comet arrives at 100,000 AU from the Alpha Centauri system, it is unlikely it will be captured into this system, because this system moves relative to our solar system by about 3.7 miles (6 kilometers) per second, which is much too fast. It would likely just fly by. In order to be captured at this distance, it would need to be going much more slowly at about .6 miles (1 kilometer) per second. In

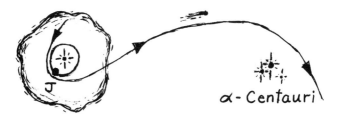

Figure 15.1 A comet passing from a chaotic orbit around the Sun, to the Oort cloud at 100,000 AU, escaping the Sun, and traveling to the Alpha Centauri system

a recent paper, I have shown together with my collaborator A. Moro-Martin that this condition could be satisfied if the comets were passing between stars in open star clusters. These are aggregates of stars lying within our galaxy. The stars in clusters are moving at about 1 kilometer per second with respect to one another, which is more desirable. So, in open star clusters, it would be possible that solar system material, such as comets, could be transferred from one system and captured into another. This implies the possibility that life-bearing material, which may be contained within comets, could be transferred between planetary systems, with a key to the origin of life within our own solar system.

Chapter Sixteen

A Paradigm Shift and the Future

. . . .

About one century ago, brothers Wilbur and Orville Wright had an idea, far-fetched at the time, that a person could fly in a vehicle heavier than air. Most people did not take this seriously and laughed. I can only imagine the ridicule. Then, on December 17, 1903, the brothers proved them wrong by demonstrating sustained flight for twelve seconds and for 120 feet at Kitty Hawk, North Carolina. The plane that did this is called the *Wright Flyer*. It wasn't in the air a long time, but it was an event that shook the world. Soon thereafter, the appearance of the plane not only transformed warfare, but the world in which we live.

The Wright Brothers shattered a paradigm—that human beings would never fly. They had an idea of how to build a heavier-than-air vehicle and give it lift. From this idea a new way of thinking arose about the capabilities of human beings. We were no longer constrained to live on the ground. The sky was now the limit. The idea that human beings could travel in space, beyond the Earth, was now a real possibility.

It was only sixty-six years later that the mighty *Saturn V* moon rocket thundered off of the launchpad at Cape Canaveral and brought *Apollo* astronauts to the Moon in the famous *Apollo 11* mission. The historic words that were spoken from the surface of the Moon on July 20, 1969, by Neil Armstrong are forever etched onto the history of the human race, "The Eagle Has Landed." The human race is now poised to go to the planet Mars.

The idea that you could use the unpredictable nature of chaos to guide a spacecraft to the Moon and have it achieve orbit with no fuel was also the shattering of a paradigm. At first this idea was viewed as far-fetched; using the high energy Hohmann transfer was the accepted norm. It was fast and easy to understand. Even though it was a fuel hog and somewhat risky, it got the job done. It was exclusively used by the two major superpowers, the United States and the Soviet Union, from the 1950s on for all their space missions.

The idea that you could gently ride the chaotic stability boundaries of gravitational fields of planets to achieve useful and fuel-efficient paths for spacecraft was not taken seriously. The fact that it took *LGAS* two years to reach the Moon was viewed as too long to be taken seriously.

The arrival by *Hiten* at the Moon on October 2, 1991, represents a paradigm shift in our perceptions of how space travel can be done. It was not reported in a major way in the press, and was known at the time only to the specialists.

But gradually the significance of *Hiten* is becoming more recognized. The use of chaos to guide spacecraft in space for less fuel is gaining more acceptance. To underscore this, Europe's *SMART 1* spacecraft arrived ballistically into orbit around the Moon in November 2004 after a two-year journey, modeled after the *LGAS* mission design. As trumpeted in the press, "*SMART 1* successfully reached the no-man's-land near the Moon, the so-called weak stability boundary, with the fuel efficiency of 4 million miles per gallon."

The proposed U.S. return to the Moon to establish a lunar base will likely make significant use of the type of lunar transfer employed by *Hiten*. Even if astronauts continue to use the old-fashioned, fuel-hungry method of the *Apollo* days to get to the Moon quickly, their tons of supplies can use the more leisurely route at a fraction of the cost.

Like the *Wright Flyer*, *Hiten* was just the beginning. Our perceptions of how to guide spacecraft are now forever changed. What lies ahead is as endless as space itself.

end

Bibliography. . . . . . .

[1] M. Adler. "To the Planets on a Shoestring." *Nature*, pp. 510–512, November 30, 2000.

[2] V. M. Alekseev. "Quasirandom Dynamical Systems." i, ii, iii. *Math. USSR Sbornik*, 5, 6, 7:73–128, 505–560, 1–43, 1960, 1960, 1969.

[3] V. I. Arnold. *Mathematical Methods of Classical Mechanics* (Springer-Verlag: New York, Heidelberg, Berlin), 1978.

[4] R. R. Bate; D. D. Mueler; J. E. White. *Fundamentals of Astrodynamics* (Dover: New York), 1971.

[5] E. Belbruno. "Through the Fuzzy Boundary: A New Route to the Moon." *Planetary Report*, 7:8–10, May/June 1992.

[6] E. Belbruno. *Capture Dynamics and Chaotic Motions in Celestial Mechanics* (Princeton University Press: Princeton), 2004. (This is the first textbook on the subject of low-energy transfers using ballistic capture presented in a rigorous fashion.)

[7] E. Belbruno. "A Low-Energy Lunar Transportation System Using Chaotic Dynamics." *Advances in the Astronautical Sciences*, 123: 2059–2066, 2006.

[8] E. Belbruno and J. R. Gott III. "Where Did the Moon Come From?" *Astronomical Journal*, March 2005.

[9] E. Belbruno and B. Marsden. "Resonance-Hopping in Comets." *Astronomical Journal*, 113:1433–1444, April 1997.

[10] E. Belbruno and A. Moro-Martin. "Slow Chaotic Transfer of Remnants Between Planetary Systems." Submitted for publication, November 2006.

[11] E. A. Belbruno. "Examples of the Nonlinear Dynamics of Ballistic Capture and Escape in the Earth-Moon System." In *Proceedings of the Annual AIAA Astrodynamics Conference*, number AAS 90–2896, August 1990.

[12] E. A. Belbruno. *Resonant Hopping in the Kuiper Belt*, vol. 522, *Series C. Mathematical and Physical Sciences*, pp. 37–49, 1997.

[13] E. A. Belbruno. "Procedure for Generating Operational Ballistic Capture Transfer Using Computer Implemented Process." Technical Report Patent No. 6,278,946, United States Patent Office, August 21 2001.

[14] E. A. Belbruno and J. Miller. "A Ballistic Lunar Capture Trajectory for the Japanese Spacecraft *Hiten*." Technical Report JPL-IOM 312/90.4-1731-EAB, Jet Propulsion Laboratory, June 15, 1990.

[15] M. Bello-Mora, F. Graziani, P. Teofilatto, C. Circi, M. Porfilio, and M. Hechler. "A Systematic Analysis on Weak Stability Boundary Transfers to the Moon." In *Proceedings of 51st Inter. Astronautical Congress*, number IAF-00-A.6.03, Rio de Janeiro, Brazil, October 2000.

[16] A.G.W. Cameron and W. R. Ward. "The Origin of the Moon." In *Proc. Lunar Planet. Sci. Conf. 7th*, pp. 120–122, 1976.

[17] R. Canup. "Simulations of a Late-Forming Impact." *Icarus*, 168:433–456, 2004.

[18] R. Canup and E. Asphaug. "Origin of the Moon in a Giant Impact Near the End of Earth's Formation." *Nature*, 412:708–712, 2001.

[19] M. Chown. "The Planet that Stalked the Earth." *New Scientist*, pp. 26–31 (cover story), August 14, 2004.

[20] C. Conley. "Low-Energy Transit Orbits in the Restricted Three-Body Problem." *SIAM J. Appl. Math.*, 16:732–746, 1968.

[21] F. Diacu. "The Slingshot Effect of Celestial Bodies." π *in the Sky*, pp. 16–17, December 2000.

[22] F. Diacu and P. Holmes. *Celestial Encounters: The Origins of Chaos and Stability* (Princeton University Press: Princeton, NJ), 1996.

[23] D. W. Dunham and R. W. Farquhar. "Background and Application of Astrodynamics for Space Missions of the Johns Hopkins Applied Physics Laboratory." In *Astrodynamics, Space Missions, and Chaos*, vol. 1017, Annals of the New York Academy of Science, pp. 267–307, May 2004.

[24] L. Dye. "Cluster Probes Look for Lift on Space Guns." *Los Angeles Times*, p. 1, January 11, 1988.

[25] L. Dye. "With a Boost from JPL, Japanese Lunar Mission May Get Back on Track." *Los Angeles Times*, science section, July 16, 1990.

[26] R. W. Farquhar. "The Control and Use of Libration-Point Satellites." Technical Report TR R-346, NASA, September 1970.

[27] R. W. Farquhar, D. P. Muhonen, C. R. Newman, and H. S. Heuberger. "Trajectories and Orbital Maneuvers for the First Libration-Point Satellite." *J. Guid. and Control*, 3:549–554, 1980.

[28] A. Frank. "Gravity's Rim: Riding Chaos to the Moon." *Discover*, pp. 74–49, September 1994.

[29] W. Gibbs. "Banzai!" *Scientific American*, p. 22, July 1993.

[30] J. Gleick. *Chaos: Making a New Science* (Viking Press: New York), 1987.

[31] W. K. Hartmann and D.R. Davis. "Satellite-Sized Planetesimals and Lunar Origin." *Icarus*, 24:504–515, 1975.

[32] D. Heggie and P. Hut. *The Gravitational Million-Body Problem* (Cambridge University Press: Cambridge), 2003.

[33] W. Hohmann. *Die Erreichbarkeit der Himmelskorper* (Oldenbourg: Munich), 1925.

[34] C. Howell, B. Barden, and M. Lo. "Application of Dynamical Systems Theory to Trajectory Design for a Libration Point Mission." *Journal of Astronautical Sciences*, 45:161–178, 1997.

[35] N. Ishii, H. Matsuo, H. Yamakawa, J. Kawaguchi. "On Earth-Moon Transfer Trajectory with Gravitational Capture." In *Proceedings AAS/AIAA Astrodynamics Specialists Conf.*, no. AAS 93-633, August 1993.

[36] D. C. Jewitt and J. X. Luu. *I.A.U. Circular* (5611), 1992.

[37] J. Kawaguchi, H. Yamakawa, T. Uesugi, and H. Matsuo. "On Making Use of Lunar and Solar Gravity Assists in Lunar A, Planet B Missions." *Acta. Atsr.*, 35:633–642, 1995.

[38] E. Klarreich, "Navigating Celestial Currents." *Science News*, pp. 250–252 (cover story), April 25, 2005.

[39] W. S. Koon, M. W. Lo, J. E. Marsden, and S. D. Ross. "Heteroclinic Connections between Periodic Orbits and Resonance Transitions in Celestial Mechanics." *Chaos*, 10(2):427–469, June 2000.

[40] B. G. Marsden. "The Orbit and Ephemeris of Periodic Comet Oterma." *The Astronomical Journal*, 66:246–248, June 1961.

[41] J. Meeus. *Astronomical Algorithms* (Willmann-Bell: Richmond), 1991.

[42] W. Mendell. "A Gateway for Human Exploration of Space? The Weak Stability Boundary." *Space Policy*, 17:13–17, 2001. (This offers an interesting perspective on weak stability boundaries.)

[43] J. Moser. *Stable and Random Motions* (Princeton University Press: Princeton, NJ), 1973.

[44] K. Nock and R. P. Salazar. "To the Moon on Gossamer Wings." *Aerospace America*, 27:40–42, March 1987.

[45] C. Ocampo. "Trajectory Analysis for Lunar Flyby Rescue Of Asiasat-3/Hgs-1." In *New Trends in Astrodynamics and Applications*, vol. 1065, Annals of the New York Academy of Sciences, pp. 232–253, December 2005.

[46] H. Pollard. *Celestial Mechanics* (Mathematical Association of America: Washington, DC), 1976.

[47] G. Racca. "New Challanges to Trajectory Design by the Use of Electric Propulsion and Other Means of Wandering in the Solar System." *Celestial Mechanics and Dynamical Astronomy*, 85:1–24, 2003.

[48] R. Ricvance. *Van Gogh in Saint-Rémy and Auvers* (Metropolitan Museum of Art, New York), 1986.

[49] C. Simó, G. Gomez, J. Llibre, R. Martinez, and J. Rodriguez. "On the Optimal Station Keeping Control of Halo Orbits." *Acta Astronautica*, 15:391–397, 1987.

[50] K. A. Sitnikov. "Existence of Oscillating Motion for the Three-Body Problem." *Dokl. Akad. Nauk USSR*, 133(2):303–306, 1960.

[51] T. Sweetser et al. "Trajectory Design for a Europa Orbiter Mission: A Plethora of Astrodynamic Challenges." In *Proceedings of AAS/AIAA Space Flight Mechanics Meeting*, no. AAS 97–174, February 1997.

[52] K. Uesugi et al. "Mission Operations of the Spacecraft Hiten." In *Proceedings of the 3rd International Symposium on Space Flight Dynamics*, Darmstadt, Germany, September 30–October 4, 1991.

[53] R. Vanderbei, "Horsing Around on Saturn." In *New Trends in Astrodynamics and Applications*, vol. 1065, Annals of the New York Academy of Sciences, pp. 337–345, December 2005.

Index